EVOLUTION OF DIVING ADAPTATIONS
IN THE STIFFTAIL DUCKS

EVOLUTION OF DIVING ADAPTATIONS IN THE STIFFTAIL DUCKS

BY
ROBERT J. RAIKOW

UNIVERSITY OF CALIFORNIA PRESS
BERKELEY · LOS ANGELES · LONDON
1970

UNIVERSITY OF CALIFORNIA PUBLICATIONS IN ZOOLOGY
ADVISORY EDITORS: G. A. BARTHOLOMEW, J. H. CONNELL, JOHN DAVIS, C. R. GOLDMAN
CADET HAND, KENNETH NORRIS, O. P. PEARSON, R. H. ROSENBLATT, GROVER STEPHENS

Volume 94

Approved for publication April 17, 1970
Issued September 29, 1970
Price, $2.50

UNIVERSITY OF CALIFORNIA PRESS
BERKELEY AND LOS ANGELES
CALIFORNIA

◇

UNIVERSITY OF CALIFORNIA PRESS, LTD.
LONDON, ENGLAND

ISBN: 0-520-09363-1
LIBRARY OF CONGRESS CATALOG CARD No.: 76-631462

© 1970 BY THE REGENTS OF THE UNIVERSITY OF CALIFORNIA
PRINTED IN THE UNITED STATES OF AMERICA

CONTENTS

List of Abbreviations of Muscles	vi
Introduction	1
Materials and Methods	2
Aquatic Locomotion	2
Position of the hind limb in swimming and diving	2
Summary of locomotor habits	4
Discussion	6
The Tail	6
Use of the tail	6
Osteology of the tail	6
Muscles of the tail	7
Discussion	14
The Hind Limb	16
The pelvic girdle	16
Osteology of the hind limb	17
Muscles of the hind limb	21
Discussion	39
General Conclusions	48
Summary	50
Acknowledgments	50
Literature Cited	51

ABBREVIATIONS OF MUSCLES USED IN FIGURES

abd. dig. II	abductor digiti II
abd. dig. IV	abductor digiti IV
add. dig. II	adductor digiti II
add. p. ext.	adductor longus et brevis pars externa
add. p. int.	adductor longus et brevis pars interna
add. rect.	adductores rectrices
bic. fem.	biceps femoris
depress. caudae	depressor caudae
depress. coccygis	depressor coccygis
ext. brev. d. IV	extensor brevis digiti IV
ext. dig. l.	extensor digitorum longus
ext. hal. l.	extensor hallucis longus
ext. pro. d. III	extensor proprius digiti III
f. dig. l.	flexor digitorum longus
fem. tib. ext. et med.	femorotibialis externus et medius
fem. tib. int.	femorotibialis internus
f. hal. brev.	flexor hallucis brevis
f. hal. l.	flexor hallucis longus
f. p. d. II	flexor perforatus digiti II
f. p. d. III	flexor perforatus digiti III
f. p. d. IV	flexor perforatus digiti IV
f. p. et p. d. II	flexor perforans et perforatus digiti II
f. p. et p. d. III	flexor perforans et perforatus digiti III
gas. p. ext.	gastrocnemius pars externa
gas. p. int.	gastrocnemius pars interna
gas. p. med.	gastrocnemius pars media
glut. med. et min.	gluteus medius et minimus
il. tib.	iliotibialis
il. troc. ant. et med.	iliotrochantericus anterior et medius
il. troc. post.	iliotrochantericus posterior
isch. fem.	ischiofemoralis
lat. caudae	lateralis caudae
lat. coccygis	lateralis coccygis
lev. caudae	levator caudae
lev. cloacae	levator cloacae
lev. coccygis	levator coccygis
obt. ext.	obturator externus
obt. int.	obturator internus
per. brev.	peroneus brevis
per. long.	peroneus longus
pirif. p. il. fem.	piriformis pars iliofemoralis
pirif. p. caud. fem.	piriformis pars caudofemoralis
sar.	sartorius
semim.	semimembranosus
semit.	semitendinosus
tib. ant.	tibialis anterior
trans.	transversoanalis

EVOLUTION OF DIVING ADAPTATIONS IN THE STIFFTAIL DUCKS

BY

ROBERT J. RAIKOW

(A contribution from the Museum of Vertebrate Zoology of the University of California)

INTRODUCTION

THE ORDER Anseriformes is an ancient group of birds, which probably arose in the Cretaceous and was well established by the Eocene. Early forms are of uncertain character, but by mid-Tertiary definite ducks, geese, swans, and whistling ducks are distinguishable. Apparently the group arose in Europe (Howard, 1950). As with birds generally, the fossil record is rather poor and sheds little light on phylogeny, hence most evolutionary studies rely on a comparison of various characters in the living forms.

Wetmore (1960) recognizes two suborders, the Anhimae and Anseres. The screamers (Anhimae) of South America are somewhat intermediate between waterfowl and the Galliformes, being only slightly aquatic in their habits and adaptations. The waterfowl (Anseres) may be described as moderately-sized, web-footed, more-or-less aquatic birds with a tufted oil gland; having downy precocial young; and having a desmognathous skull with basipterygoid facets and a lamellate bill. There are many forms, but the distinctions between them are not profound, based mainly on plumage, bodily proportions, and behavior.

In this monograph the classification of Johnsgard (1968) will be followed. His representation of the phylogeny of the waterfowl is based on the work of Delacour and Mayr (1945). In this scheme the true ducks comprise the subfamily Anatinae of the family Anatidae, the Anatinae being further subdivided into tribes of from one to nine genera each. It is believed that the several tribes specialized for diving originated independently from a surface-feeding, non-diving ancestral stock represented today by the tribe Anatini, the dabbling ducks. One of these diving groups is the tribe Oxyurini, the stifftails, including the genera *Heteronetta*, *Oxyura*, and *Biziura*. The present study is concerned with an analysis of the evolution of diving adaptations in the stifftails, especially the osteology and myology of the hind limb and tail. It is based on dissection of three species of Oxyurini (*Heteronetta atricapilla, Oxyura jamaicensis,* and *Biziura lobata*) which represent a presumed evolutionary sequence, and one species of Anatini (*Anas platyrhynchos*) which represents the form of the ancestral surface-feeding stock from which the Oxyurini presumably arose. In the text these will usually be referred to only by the generic name, but unless stated otherwise such references will pertain only to the species named.

The aims of this study are: to trace the evolutionary history of the morphological modifications associated with the change from a non-diving to a diving mode of life, to analyze the adaptive and mechanical significance of these changes, and to clarify the taxonomic relationships among the genera under investigation.

MATERIALS AND METHODS

Observations and measurements were made on prepared skeletons of ducks from the collections of the Museum of Vertebrate Zoology, University of California, Berkeley, and the University of Michigan Museum of Zoology. Measurements were made to the nearest $1/10$ mm with a dial caliper.

I dissected six specimens of the Mallard (*Anas platyrhynchos*), five of the Ruddy Duck (*Oxyura jamaicensis*), and one each of the Black-Headed Duck (*Heteronetta atricapilla*) and Musk Duck (*Biziura lobata*). In addition several specimens of the Green-winged Teal (*Anas crecca*) were examined for comparison with the Mallard, but they will not be discussed here. *Anas* and *Oxyura* were collected in the vicinity of Berkeley, California, while *Heteronetta* and *Biziura* were obtained from the collection of the Peabody Museum of Natural History, Yale University. Dissection of smaller structures was done with the aid of a binocular dissecting microscope under 10× and 20× magnification.

The dry weights of muscles were obtained by removing the muscle from the preserved specimen, cleaning off all superficial connective tissue, nerves, etc., and removing external and internal tendons. The muscle was then pressed between pieces of absorbent paper to remove much of the fluid, and finally dried for a minimum of 48 hours (usually 72 hours) in a drying oven at 105°C. The percent weight of a given muscle is hereby defined as the percent which the dry weight of the muscle comprises of the dry weight of the sum of all the muscles of the limb.

Drawings were made freehand with the aid of proportional dividers. Locomotion in Mallards and Ruddy Ducks was observed at Lake Merritt, Oakland, California, and at the Marina in Berkeley, California.

Quantitative data given in several tables were tested for statistical significance using the Student's t-test.

AQUATIC LOCOMOTION

Position of the Hind Limb in Swimming and Diving

Stolpe (1932) noted that many aquatic birds have two regular positions of the hind limb, one used in walking and the other in swimming. In walking the leg is strongly adducted and directed downward so that the tarsus is nearly vertical in orientation, lying in a parasagittal plane, and the tibia is also greatly extended. The toes are spread on the ground in the horizontal plane. In swimming, however, the leg is abducted and the tibia is held in a more horizontal, flexed position and directed medially, so that the two tibiae converge posteriorly. The tarsi are abducted so that they diverge rather than lie parallel to each other. At its extreme, this position results in the femora and tarsi being abducted to the extent that they lie in or near the horizontal plane. This extreme abduction is seen in loons and grebes, where the shank is so tightly bound into the body musculature and held there by the skin, that extension of the tibia is prevented and the bird permanently holds this abducted posture of the leg, which makes walking on land extremely difficult. The adducted, walking posture is illustrated

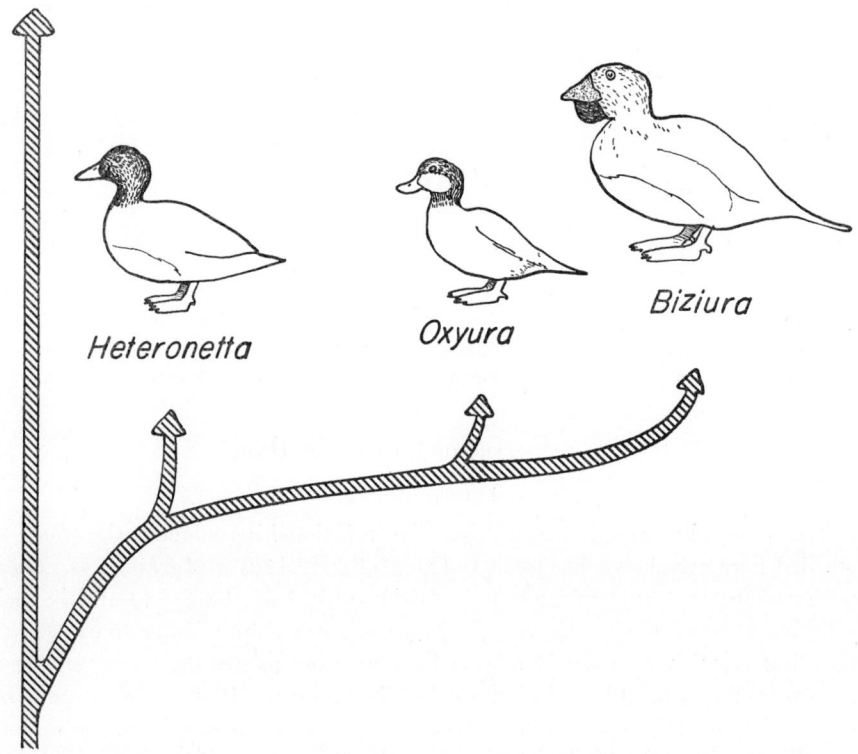

Fig. 1. The phylogenetic relationships among *Anas platyrhynchos*, representing the tribe Anatini, and the three genera of the tribe Oxyurini. The species are *Heteronetta atricapilla*, *Oxyura jamaicensis*, and *Biziura lobata*.

by the ducks in figure 1, while the abducted, diving posture is diagrammed in fiigure 2.

A bird may swim either with its legs in the abducted position, or by holding them in the walking position and "dog-paddling." The former is much more efficient, as explained by Dabelow (1925) and is utilized by the most highly specialized divers, including loons and grebes. Mediocre swimmers, such as gulls, use the latter method.

Detailed studies of the underwater movements of ducks have not been reported, but some generalizations may be made on the basis of my observations of Mallards and Ruddy Ducks and information from the literature.

Fig. 2. Diagram showing the abducted position of the hind limbs in a diving duck. The wings and tail have been omitted for clarity. The adducted position used in walking is shown in figure 1.

SUMMARY OF LOCOMOTOR HABITS

Anas platyrhynchos

The generally accepted phylogenetic relationships among the four species studied here are shown in figure 1. The Mallard, *Anas platyrhynchos*, was chosen to represent the Anatini since it is fairly typical of the group and is easily obtainable. It is relatively unspecialized in locomotor habits, being an excellent swimmer, but is neither as highly adapted for walking as are the geese nor for diving as the stifftails or other diving ducks. Weidmann (1956) has summarized the locomotion of this species. The legs are short, resulting in the characteristic waddling gait as the bird pivots with each step to keep its center of gravity over the foot on the ground. The legs are sufficiently far forward, however, so that the bird can maintain its balance along its longitudinal axis. In slow swimming the femur and tibiotarsus remain still, and the power stroke is solely by means of extension of the tarsometatarsus. In rapid swimming, however, this movement is augmented by a piston-like back-and-forth action of the tibiotarsus effected by alternate extension and flexion of the femur. The toes are spread in the backstroke and folded together during the recovery stroke. The Mallard ordinarily does not dive, except when being pursued.

Heteronetta atricapilla

The Black-Headed Duck, *Heteronetta atricapilla*, of South America, is generally regarded as being the least specialized of the Oxyurini, occupying an intermediate position between the Anatini and the more specialized Oxyurini

(Delacour, 1959:262). The bill and feeding habits are dabbler-like (Weller, 1968:176). Osteologically it is less specialized than other Oxyurini (Woolfenden, 1961:122). The plumage of the downy young resembles that of Anatini (Johnsgard, 1965:318), while the juvenal plumage is most like *Oxyura* (Weller, 1967: 136). In sexual behavior it shares characteristics of both dabbling and stifftail ducks, resembling the latter more closely (Johnsgard, 1965:320).

The habits of the Black-Headed Duck have not been widely studied, but Weller (1968) has recently provided a summary of its locomotor behavior. It is more aquatic than dabblers, and only rarely comes ashore to sleep. It walks poorly but may stand in shallow water to preen. When swimming its profile resembles that of divers, with the upper back being higher than the tail, which may either be held upraised, or lowered to the surface of the water, as in the Ruddy Duck. It is reluctant to fly, especially during the day, but can fly very fast with rapid wingbeats, and rises directly from the water like dabbling ducks, rather than running along the surface to take off like typical stifftails. This is considered an adaptation to life in small pools. It dives skillfully to obtain food, but does not use its wings underwater as postulated by Wetmore (1926), and apparently cannot submerge gradually as do the Ruddy Duck and Musk Duck. It lacks the lobed hallux of the Ruddy and Musk Ducks; this is generally regarded as an aquatic adaptation and is found in most diving ducks.

Oxyura jamaicensis

The genus *Oxyura* includes six species and has a worldwide distribution. Many workers consider *Oxyura jamaicensis* and *O. ferruginea* conspecific, but they are listed separately in figure 3 because the specimens measured were so labeled. Except for the diminutive Masked Duck (*O. dominica*), which is sometimes placed in a separate genus (*Nomonyx*), the species are quite similar in most respects and in this study only the North American Ruddy Duck, *O. jamaicensis* has been considered in any detail. According to Kortright (1942) the Ruddy Duck cannot walk on land for more than a few steps without falling forward onto its breast. It is strictly aquatic and the extreme posterior position of the feet, which makes walking so difficult, adapts it excellently for swimming. The Ruddy Duck swims low in the water with the tail either raised up sharply or held flat against the water's surface. It may dive either with a jump or by submerging gradually and quietly, and it swims underwater with both feet paddling simultaneously and the wings closed. When taking off it must run for some distance along the surface (Delacour, 1959:223) but flies rapidly. As in the other genera of Oxyurini, Ruddy Ducks seldom fly during the day.

The habits of other species of *Oxyura* are similar except that *O. dominica* can take off by leaping upward rather than having to run along the surface (Weller, 1968). It is smaller than the other species and probably has a lower wing loading.

Biziura lobata

The most highly specialized of the Oxyurini is the bizarre Musk Duck of Australia. This species is noted for the peculiar black lobe hanging beneath the bill of the male and the extreme sexual dimorphism in size, with the males weighing two to three times as much as the females (Johnsgard, 1966:98). Frith (1967) notes that the Musk Duck almost never comes to land and has difficulty standing, but must slither along on its belly. As in the Ruddy Duck, this is because of the extreme posterior position of the feet. It dives well and can sink slowly beneath the water like the Ruddy Duck. A long distance is required for takeoff, landing is clumsy, and it flies mostly at night.

DISCUSSION

To what extent do the swimming postures of ducks approach the extreme abducted position of the most highly specialized divers? When swimming on the surface the Mallard holds the tarsi slightly abducted, perhaps 45° from the vertical at most. The tarsi of the Ruddy Duck appear to be abducted somewhat more than this. Excellent photographs of the Blue-Billed Duck (*Oxyura australis*) and Musk Duck swimming underwater are provided by Frith (1967: opposite pp. 298 and 299). In each case the legs are highly abducted, reaching or possibly exceeding the horizontal plane, with the feet converging medially at the posterior end of the body. Undoubtedly the Ruddy Duck utilizes a comparable movement. Examination of dead specimens of these species reveals that in the "relaxed" posture of death the hind limb of *Oxyura* and *Biziura* is much more abducted than that of *Anas*, and analysis of the skeleton (see below) indicates that the pelvic girdle and hind limb are modified to enhance the efficiency of this posture. In *Heteronetta* the limb position resembles that of *Anas* rather than *Oxyura*. Thus it may be concluded that in the evolution of the diving habit in the stifftail ducks there has been a change in the position of the hind limbs correlated with the adoption of a more efficient mode of underwater progression.

THE TAIL

USE OF THE TAIL

The tail has three important functions in Oxyurine ducks. It is used in flight, especially as a brake in landing; it is used in sexual displays; and it is used as a rudder underwater. The Oxyurini are commonly called "stifftail ducks" owing to the stiffened rhachises of the rectrices in all three genera. Undoubtedly this modification, which increases resistance to bending, promotes greater effectiveness when the tail is used as a rudder underwater.

OSTEOLOGY OF THE TAIL

The number of caudal vertebrae varies with age. It is highest in young birds, then decreases as elements become incorporated into the synsacrum or pygostyle (Verheyen, 1955). Woolfenden (1961) has summarized the data on number of caudal vertebrae, which include all caudals not fused to the synsacrum, as follows: *Anas platyrhynchos*, 7–8; *Heteronetta atricapilla*, 8–9; *Oxyura jamaicensis*, 8–9; and *Biziura lobata*, 7 plus. For the purpose of functional analysis, however, it is better to consider only those vertebrae freely capable of movement. Usually there are one or two anterior free caudals which are not fused with the synsacrum, but which lie between the postero-dorsal ends of the ilia, and are bound to them at least by ligaments and whose mobility is thus restricted or eliminated. For this reason I define "postpelvic caudal vertebrae" as those which lie posterior to the pelvic girdle, are not connected to it, and thus are freely movable, but not including the pygostyle. The numbers of postpelvic caudal vertebrae are given in table 1. The addition of two postpelvic caudal vertebrae in *Biziura* probably increases the flexibility of the tail in locomotion, and undoubtedly contributes to this species' ability to perform extreme tail-cocking during its "whistle-kick" display, as illustrated by Johnsgard (1966:100), in which the tail is raised so far over the back that the rectrices are drawn down horizontally and point directly anteriorly.

TABLE 1
NUMBERS OF POSTPELVIC CAUDAL VERTEBRAE IN STIFFTAIL DUCKS

Species	No. examined	No. of postpelvic caudal vertebrae
Anas platyrhynchos	4	6
Heteronetta atricapilla	2	6
Oxyura jamaicensis	2	5
	4	6
Biziura lobata	2	8

LENGTH OF THE CAUDAL VERTEBRAL COLUMN

"Free tail length" is defined as the sum of the lengths of the postpelvic vertebrae and the pygostyle. The length of the vertebrae was taken as the midventral length of the centra, while the length of the pygostyle was measured as the distance from the tip to the articular surface for the last caudal vertebra. To obtain an estimate of these values relative to the total size of the specimen, they were compared to "trunk length," which is defined as the distance from the anterior articular surface of the centrum of the most posterior cervical vertebra (as defined by Bellairs and Jenkin, 1960:249) to the posterior edge of the most posterior lumbar vertebra (as shown by Howard, 1929:321). These measurements are given in table 2.

The ratios of free tail length to trunk length are shown in figure 3. *Anas* and *Heteronetta* have the relatively shortest free tail lengths, while *Oxyura* shows a moderate elongation. *Biziura* has a greatly elongated caudal skeleton, primarily due to the addition of two vertebrae rather than by elongation of the pygostyle (table 3). These are not size-dependent changes as may be seen from tables 2 and 3. *Anas* has the shortest tail length, but is larger than *Heteronetta*. The latter, with an equally short tail length, is larger than *Oxyura*. The elongation of free tail length in *Oxyura* and *Biziura* presumably provides a greater range of movement to the tail, especially in a dorso-ventral plane. This is correlated with these species' use of the tail as an underwater rudder, as well as in sexual display.

MUSCLES OF THE TAIL

The functional morphology of the caudal musculature in birds has not been widely studied, despite the excellent review by du Toit (1913). The major exceptions are Fisher (1946) and Owre (1967), who have correlated variations in myology with the use of the tail in locomotion. The terminology used for the muscles here is that of George and Berger (1966).

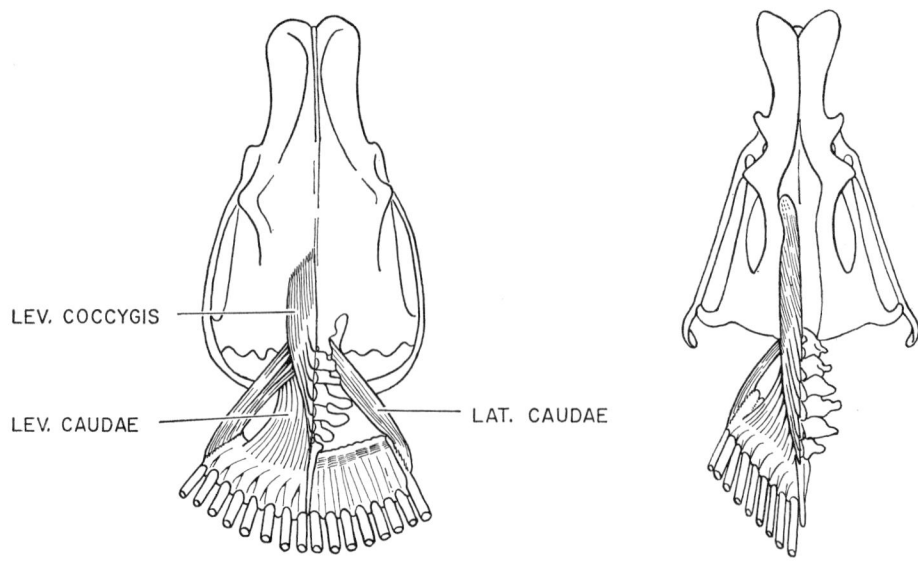

Anas   *Oxyura*

Fig. 4. A dorsal view of the caudal musculature in *Anas platyrhynchos* (left) and *Oxyura jamaicensis*. In *Anas* the lev. coccygis and lev. caudae have been removed to show the origin of M. lat. caudae on the right side.

(fig. 4). This muscle is especially massive in *Biziura* (table 4), lying deeply between the posterior iliac crest and the median dorsal ridge of the synsacrum.

M. LATERALIS CAUDAE (figs. 4, 6, 7, 8)

Structure.—A narrow, parallel-fibered muscle whose three heads arise from the transverse processes of the first three free caudal vertebrae, and the membrane joining them to the posterior edge of the ilium. The belly passes posterolaterally to insert on the lateral side of the fascial covering of the base of the outermost rectrix.

Action.—Abduction of the rectrices. The result is the spreading of the tail since the rectrices are interconnected on each side by a webbing of connective tissue as well as the fibers of Mm. adductores rectrices.

Comparison.—Extremely reduced in *Biziura*.

M. TRANSVERSOANALIS (figs. 5, 8)

Structure.—This thin, flat, parallel-fibered muscle arises on the posterior border of the ilium, ischium, and proximal one-third of the postischiadic pubis. The proximal part of the origin lies deep to the belly of M. piriformis pars caudofemoralis. The muscle passes ventrally to insert on a midline raphe anterior to the cloaca and between the ends of the pubes.

Action.—Contributes strength to the body wall, but apparently does not affect movements of the tail.

M. DEPRESSOR CAUDAE (figs. 5, 7, 8)

Structure.—This thin, flat muscle arises from the distal one-half of the postischiadic pubis and the adjacent aponeurosis medial to the end of the pubis. It passes dorsocaudally, deep to M. transversoanalis and superficial to M. depressor coccygis, and inserts on the ventral side of the tail among the bases of the three or four outer rectrices, deep to M. piriformis pars caudofemoralis.

Action.—Depresses the tail.

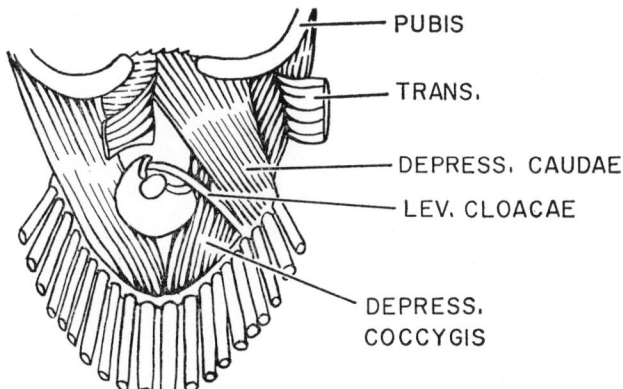

Fig. 5. Superficial ventral caudal musculature of *Anas platyrhynchos*. Mm. *transversonalis*, depressor caudae, and levator cloacae have been removed on the left side.

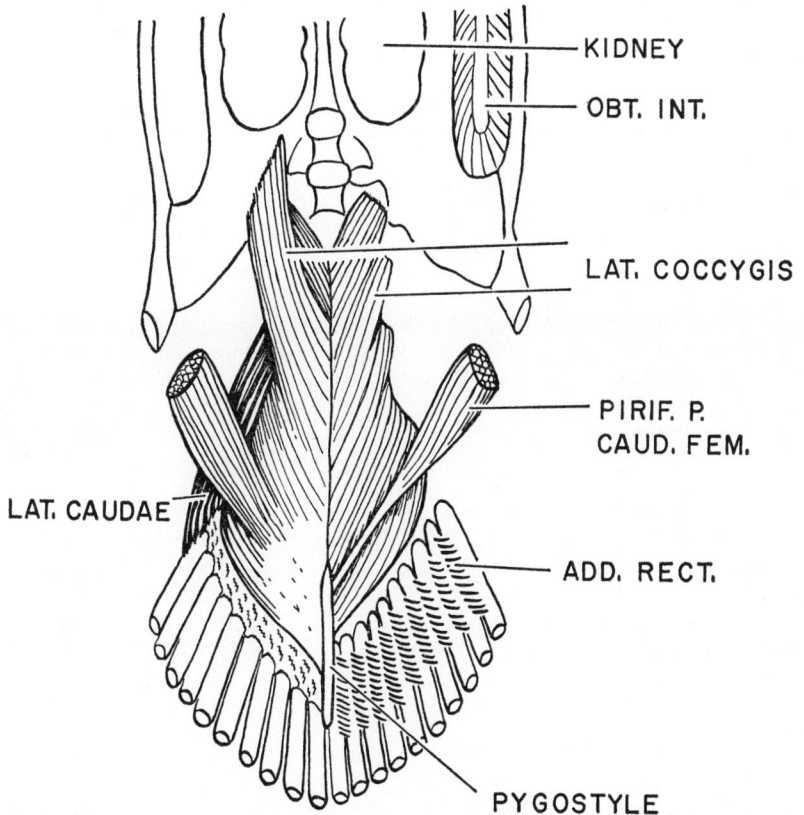

Fig. 6. Deep ventral caudal musculature of *Anas platyrhynchos*. The superficial portion of M. lat. coccygis has been removed on the right side, as has the rectricial portion of the same muscle, M. lat. caudae, and the fatty connective tissue surrounding the bases of the rectrices.

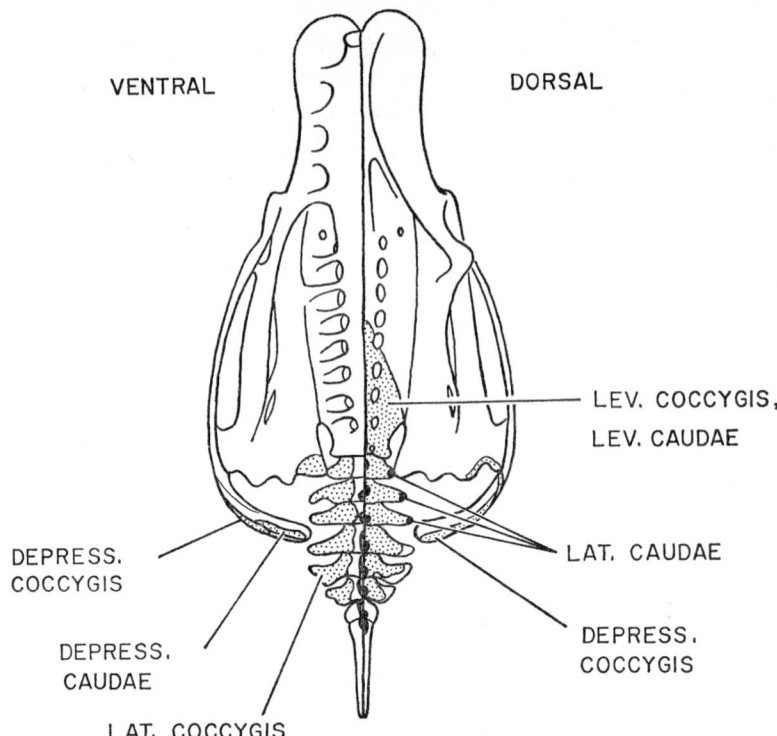

Fig. 7. Areas of attachment of caudal muscles to the pelvic girdle and caudal skeleton of *Anas platyrhynchos*.

Comparison.—In *Heteronetta* it inserts at the bases of the four or five outer rectrices, in *Oxyura* at the bases of the two or three outer rectrices. In *Biziura* the origin is more medially located, arising only from the distal tip of the pubis, and mostly from the central aponeurosis and medial edge of the same muscle of the other side. In the other species the muscles of the two sides do not quite meet.

M. LEVATOR CLOACAE (fig. 5)

Structure.—A very small, narrow, parallel-fibered muscle which arises from the medial edge of the insertion of M. depressor caudae and the adjacent fascial covering of the bases of rectrices 3 and 4. It extends ventromedially to insert on the midline in the fascial covering of the anterior side of the cloacal musculature.

Action.—Raises cloaca.

Comparison.—In *Biziura* this muscle arises more laterally, at the base of the second rectrix.

M. DEPRESSOR COCCYGIS (figs. 5, 7, 8)

Structure.—This parallel-fibered muscle has a fleshy origin from the posterior margin of the proximal two-thirds of the postischiadic pubis. It inserts on the underside of the tail in the dense connective tissue at the base of the inner five or six rectrices and the fascial covering of the underlying M. lateralis coccygis. A few fibers pass from the medial edge of the muscle to insert in the fascia around the base of the cloaca.

Action.—Depresses the tail.

Comparison.—In *Oxyura* the line of origin is slightly longer than in *Anas*, extending nearly to the distal end of the postischiadic pubis. In *Biziura* it arises from the ventral one-fourth of the posterior end of the ischium and the entire postischial pubis except the very tip.

M. LATERALIS COCCYGIS (figs. 6, 7, 8)

Structure.—This muscle can be roughly divided into a superficial and a deep portion on the basis of origin. The superficial portion arises from the ventral surface of the postero-medial end of the ilium adjacent to the synsacrum and just posterior to the kidney; and from the fascia separating it from M. lateralis caudae. The deeper portion arises segmentally from the ventral surface of the transverse process of the last fused vertebra of the synsacrum and all but the most posterior caudal vertebrae. The fibers of both portions pass postero-medially, converging from either side of the tail to fuse in the midline and forming a thick fascial covering over the posterior portion of the muscle. It inserts fleshily on the ventral surface of the centra and the hemal spines of all free caudal vertebrae, and by tendons on the apices of the hemal spines of the posterior three free caudals, and the anterior, ventral margin of the pygostyle. The most posterior portion of the muscle arises from the base of the follicle of the outer rectrix just medial to the insertion of M. lateralis caudae and has a broad, fleshy insertion on the blade of the pygostyle and the bases of the inner three or four rectrices.

Action.—Depresses the tail. The posterior portion may adduct the bases of the rectrices and thus aid M. lateralis caudae in spreading the tail. These movements occur together in the braking motion during landing.

Comparison.—The portion which arises from the rectrices is more distinctly separable from the proximal, vertebral portion in the stifftails than in *Anas*. Functionally, it is probably more important in tail spreading than in depression.

M. PIRIFORMIS PARS CAUDOFEMORALIS (figs. 6, 8, 17)

Structure.—Since both ends of this muscle are movable it is somewhat arbitrary which end is called the origin and which the insertion, but the most common usage is followed here. The origin is by a tendon from the middle of the ventral edge of the pygostyle, where it is closely associated with, but separable from the dense superficial fascia of M. lateralis coccygis. The spindle-shaped belly extends anteriorly ventral to M. piriformis pars iliofemoralis which it penetrates but with which it does not fuse. Anteriorly the belly narrows to a strong tendon which inserts in common with that of M. piriformis pars iliofemoralis on the posterior edge of the femur. A few fibers of pars iliofemoralis insert onto this tendon.

Action.—The fact that both ends are movable has caused discussion as to the function of this muscle; this has been reviewed by Owre (1967:81–82). If the tail were fixed in position, the muscle would extend the femur; but if the latter were fixed it would depress the tail, either unilaterally or bilaterally.

Comparison.—In *Heteronetta*, *Oxyura*, and *Biziura* the tendon of origin is not as distinctly separable from the dense aponeurotic covering of M. lateralis coccygis as it is in *Anas*.

MM. ADDUCTORES RECTRICES (fig. 6)

Structure.—These comprise a series of short, separate bundles of muscle fibers which interconnect the ventral side of the bases of the rectrices. The fibers are angled so as to run posteriorly from each feather to the one next lateral to it. There is also a small band of fibers on the dorsal side and oriented at right angles to the shafts of the rectrices. Another very small muscle arises on the dorsal, distal tip of the pygostyle and passes anteriorly to insert on the base of the first (innermost) rectrix. This is so rudimentary in *Anas* that it is not apparent whether any muscle fibers are present in addition to connective tissue. I cannot discern any fibers binding the first rectrix to the blade of the pygostyle; this connection is apparently entirely by connective tissue.

Action.—Adduction of the rectrices.

Comparison.—In *Oxyura* the small connection from the tip of the pygostyle to the base of the first rectrix is better developed than in *Anas*, and contains discernible muscle fibers.

MM. INTERSPINALES

Structure.—These are small, paired muscles connecting the neural spines of successive free caudal vertebrae, including the pygostyle.

Action.—Aids tail levators.

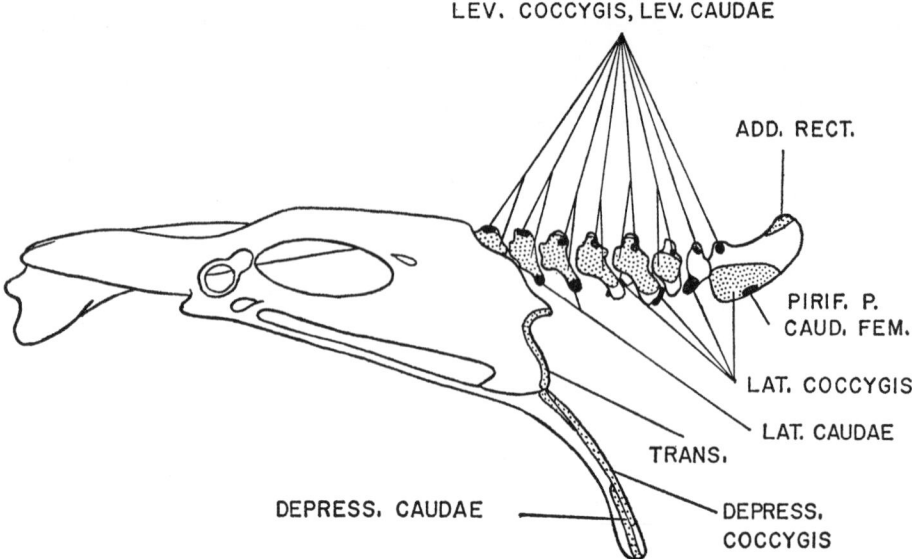

Fig. 8. Lateral view of the areas of attachment of the caudal muscles on the pelvic girdle and caudal skeleton of *Anas platyrhynchos*.

DISCUSSION

The tail consists of a set of rectrices whose bases lie side by side in a fatty, connective tissue matrix, and are joined together by connective tissue and the bands of muscle fibers collectively comprising the M. adductores rectrices. The central pair of rectrices is attached to the pygostyle, and most of the movements of the tail are accomplished by the movements of this bone. These include the following:

1. In the sagittal plane the tail is raised by bilateral action of Mm. levator coccygis and levator caudae; and lowered by Mm. lateralis coccygis, depressor caudae, depressor coccygis, and piriformis pars caudofemoralis.

2. In the horizontal plane the tail is deflected to either side by unilateral contraction of the above muscles. Tilting of the tail at any angle is accomplished by varying the contraction on one side of either levators or depressors, or of levators and depressors of opposite sides. The latter are especially well arranged to give varying degrees of depression because of their different angles of insertion onto the tail (fig. 8).

3. One movement which is apparently performed separately from any movement of the vertebral column is the spreading of the tail, which is accomplished primarily by the action of M. lateralis caudae, perhaps aided by the distal, fan-shaped portion of M. levator caudae and the posterior portion of M. lateralis coccygis. Closing of the tail is apparently the function of M. adductores rectrices. This action is probably aided by elastic recoil of the connective tissue between the follicles of the rectrices; at any rate in a dead bird the tail assumes a somewhat closed position, to which it will return if spread by hand and released.

The same muscles occur in all species studied, but there are important differences in both form and relative development. The greatest difference is in M. levator coccygis and levator caudae. In *Anas* and *Heteronetta* this muscle arises from far back on the pelvic girdle, occupying only about one-third of the length of the synsacrum. In *Oxyura* and *Biziura,* however, the area of origin is extended anteriorly almost to the level of the antitrochanter, so that it occupies about one-half the length of the synsacrum (fig. 4). This increase is associated with a relative increase in size (table 4), especially in *Biziura*. Equally important, this increase in length provides for a greater length of contraction, which must contribute to the ability of these species to cock the tail to an extreme degree. This

TABLE 4
Dry Weights of Caudal Muscles Expressed as a Percentage of the Total Dry Weight of the Caudal Muscles[a]

Muscle	Anas	Heteronetta	Oxyura	Biziura
Levator coccygis and levator caudae....	20.12 (18.55–21.69)	21.77	23.37	33.22
Lateralis caudae......................	2.68 (2.52–2.84)	1.62	1.06	0.63
Depressor caudae.....................	15.60 (14.18–17.01)	13.78	7.03	9.71
Depressor coccygis....................	18.37 (17.54–19.19)	18.63	11.36	14.71
Lateralis coccygis....................	26.31 (24.33–28.28)	30.00	32.21	27.87
Piriformis pars caudofemoralis.........	16.95 (14.17–19.72)	14.23	24.97	13.82

[a] Based on two specimens of *Anas*, and one each of the other species. The values for *Anas* are mean and range.

muscular modification therefore is associated with the increase in the length of the tail (described above in the section on The Tail) as part of an adaptive mechanism for increased mobility and strength of contraction in raising the tail.

There appears to be a trend in the phylogenetic series of genera for a relative reduction in size of M. lateralis caudae. The reason for this is not readily apparent. Tail-spreading would seem to be an important component of all species' activities; possibly it is related to flight habits rather than diving behavior. *Anas* is a much more agile flyer than the stifftails, and makes use of smaller ponds and restricted areas, into which it can drop with fair agility for a duck. It is more maneuverable in flight than the stifftails, owing in large part to a lower wing loading and different wing shape. This is a topic which I hope to investigate later. The tail is widely spread and important in hovering and landing, changing direction, etc., and the large size of M. lateralis caudae in *Anas* may be of significance in this respect.

M. piriformis pars caudofemoralis is greatly enlarged in *Oxyura*. If the depressor muscles of the tail (Mm. depressor caudae, depressor coccygis, lateralis coccygis, and piriformis pars caudofemoralis) are considered together, it is found that they account for the following percentage of total caudal muscle weight: *Anas*, 77.23; *Heteronetta*, 76.64; *Oxyura*, 75.57; *Biziura*, 66.11. It thus seems possible that enlargement of the muscle in *Oxyura* would compensate for reduction in relative size of other depressors, Mm. depressor caudae and depressor coccygis (table 4) so as to provide more power for tail depression. Similar compensation has not occurred, however, in *Biziura*.

Since these muscle weights are based on so few specimens, they should be con-

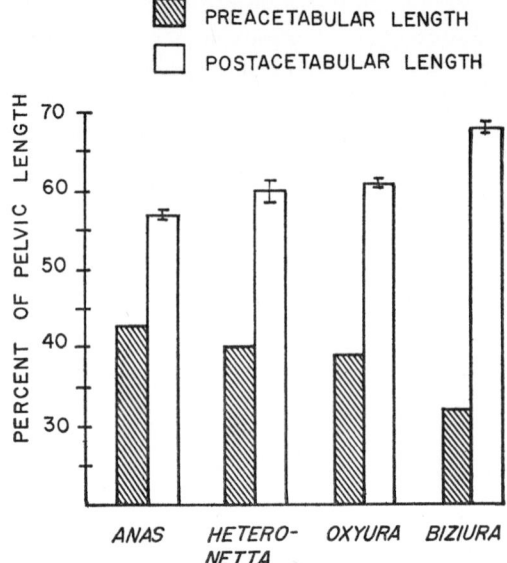

Fig. 9. Preacetabular and postacetabular length of the pelvis expressed as a percentage of total pelvic length in *Anas platyrhynchos, Heteronetta atricapilla, Oxyura jamaicensis,* and *Biziura lobata.* Vertical lines indicate two standard errors on either side of the mean. The difference between *Anas* and *Oxyura,* and between *Oxyura* and *Biziura* are statistically significant (p < .001) but the difference between *Oxyura* and *Heteronetta* is not.

sidered only as indications of general trends, but the small variation between the two specimens of *Anas* suggest that the data are valid when large interspecific differences are considered.

THE HIND LIMB

The Pelvic Girdle

Illustrations and detailed descriptions of the pelvic girdles of ducks have been given by Shufeldt (1909), Reynolds (1913), Boas (1933) and Woolfenden (1961) and need not be repeated here. In general the pelvic girdles of all ducks are similar in form, and differences are mainly in terms of relative proportions.

POSTACETABULAR EXTENSION

In aquatic birds the postacetabular part of the pelvis is relatively long compared to the preacetabular part (Dabelow, 1925). According to Miller (1937) this is associated in geese with "large size and more powerful leverage of the M. biceps femoris and M. semitendinosus." I measured the distance from the anterior tip of the ilium to the center of the acetabulum (preacetabular length) and from this point to the posterior end of the ischium (postacetabular length). The sum of these gives the total length of the pelvis. It is clear from figure 9 that in the evolution of the Oxyurini there has been a trend for elongation of the postacetabular pelvis. The difference between *Oxyura* and *Heteronetta* is not statistically significant, but both are significantly elongated compared to *Anas,* (P < .001) and the same is true for *Biziura* compared to *Oxyura* and *Heteronetta* (P < .001).

I made similar measurements of specimens representing most of the genera of ducks, and the elongation shown by *Biziura* (68 percent of total) exceeds that in any other genus, and even exceeds that in a loon (*Gavia adamsi,* 64 percent) and a grebe (*Aechmophorous occidentalis,* 63 percent). Pycraft (1906) noted the unusual proportions of the pelvis of *Biziura* but did not give a functional explanation for their relationship to the diving habits of the species.

WIDTH OF THE PELVIS

It is characteristic of diving birds that the pelvis is rather narrow (Dabelow, 1925). There is no need for a wide separation of the legs since these birds seldom walk on land, and this narrowness permits a better streamlining of the body, giving more of a teardrop shape as the bulk of the leg muscles merges smoothly with the body musculature. Such streamlining is important in species that swim underwater since the amount of resistance to the surrounding water is proportional to the cross-sectional area of the body of the bird.

The width of the pelvic girdle was measured at three points and compared to pelvic length (see figs. 10 and 11). The maximum width of the preacetabular ilium shows a slight decrease in the Oxyurini as compared to *Anas.* The interacetabular width, measured as the distance between the dorsal edges of the acetabula shows a progressive narrowing in the sequence *Anas, Heteronetta, Oxyura, Biziura.* This is the greatest degree of narrowing of any part of the pelvic girdle. It permits the heads of the femora to arise more medially from the pelvis, thus limiting the extent to which the limbs project laterally, and reducing the maximum cross-sectional area of the bird so as to minimize water resistance. Across the posterior ends of the ischia the sequence *Anas, Heteronetta, Biziura* shows a slight tendency for narrowing, but *Oxyura* has the broadest dimension here of all. Compared to a loon (*Gavia adamsi,* 40 percent of pelvic length) and especially a grebe (*Aechmophorous occidentalis,* 13 percent) this is relatively slight modification. It is associated with the fact that in loons and grebes the caudal musculature is much reduced, and steering is apparently done with the feet, while in stifftail ducks the tail is used as a rudder. Hence the caudal muscles are well developed, and require a fair amount of lateral displacement in their origins on the ischium and pubis so as to have space to permit a significant amount of deflection of the tail to each side.

OSTEOLOGY OF THE HIND LIMB

RELATIVE PROPORTIONS OF THE HIND LIMB

The sequence of genera *Anas, Heteronetta, Biziura* shows a slight tendency for an increase in the length of the tibiotarsus, measured between the proximal and distal articular surfaces (fig. 12), but this is not seen in *Oxyura.* In the sequence *Anas, Heteronetta, Oxyura, Biziura* there is a small but progressive trend toward shortening of the tarsometatarsus. Together these changes tend to place the foot more posteriorly and to draw the paddle closer to the body. This presumably increases the efficiency of the swimming movements described above in the Aquatic Locomotion section. Placement of the foot more posteriorly reduces the degree to which the body interferes with the convergent movement of the paddles. Short-

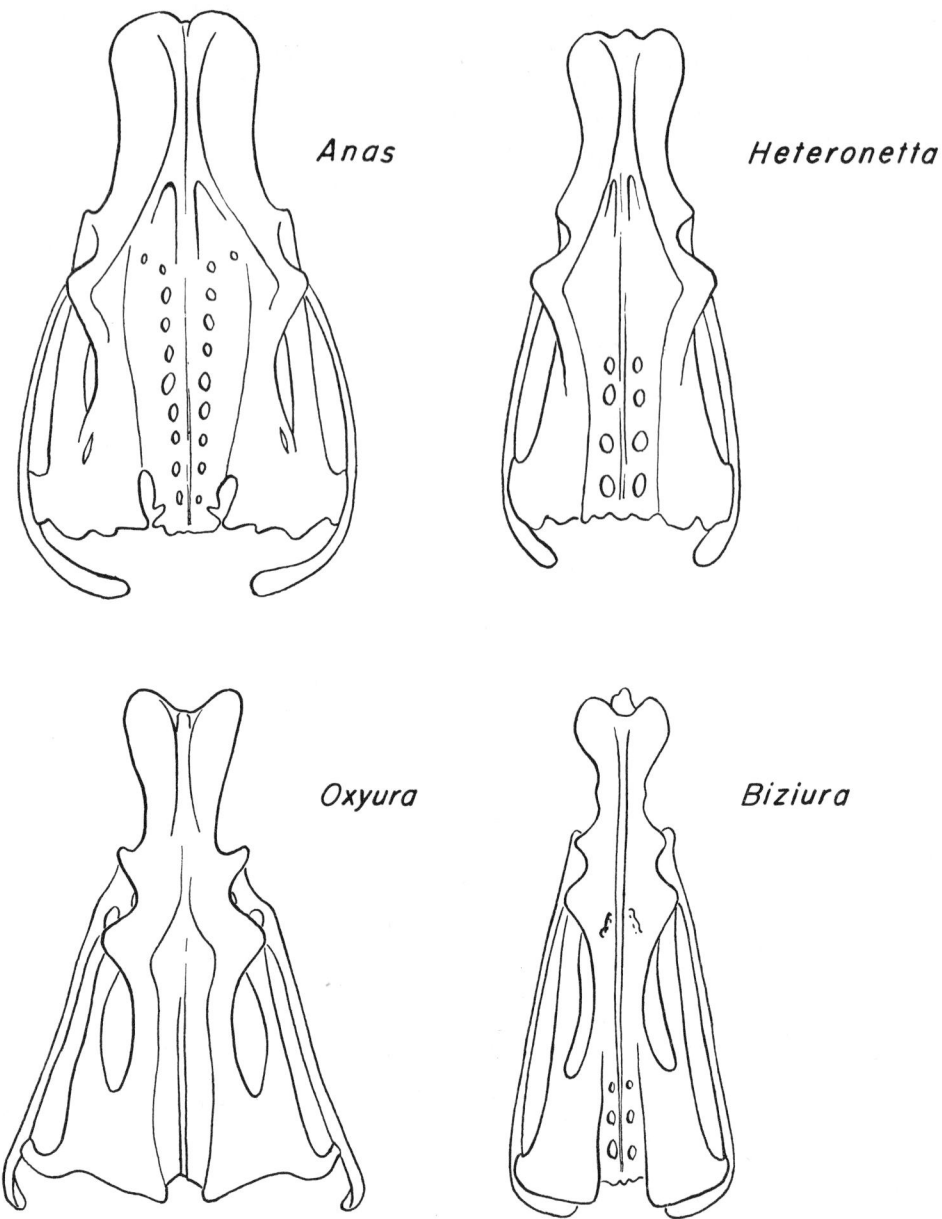

Fig. 10. Dorsal views of the pelvic girdle in *Anas platyrhynchos, Heteronetta atricapilla, Oxyura jamaicensis,* and *Biziura lobata.*

ening of the tarsometatarsus reduces the arc through which the paddle must move in displacing a given amount of water backwards during the power stroke, its shortened resistance arm giving it a higher mechanical advantage (see below, p. 39).

A much greater change is seen in the elongation of the digits. This is clearly of value in aquatic locomotion since it permits an increased area of webbing be-

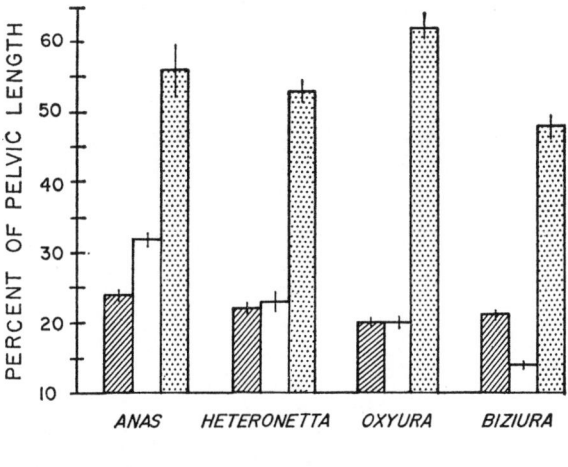

Fig. 11. Width of the pelvic girdle at three points expressed as a percentage of the length of the pelvis in *Anas platyrhynchos*, *Heteronetta atricapilla*, *Oxyura jamaicensis*, and *Biziura lobata*. The vertical lines at the top of the columns indicate two standard errors on either side of the mean.

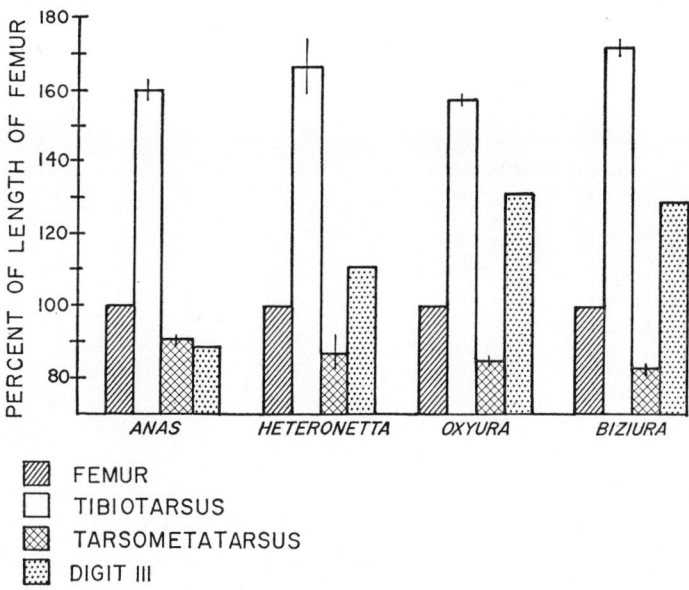

Fig. 12. Relative proportions of the hind limb in four species of ducks, *Anas platyrhynchos*, *Heteronetta atricapilla*, *Oxyura jamaicensis*, and *Biziura lobata*. The difference in length of the tibiotarsus is statistically significant in *Biziura* compared ($p < .001$) to *Oxyura* or *Anas* but not in *Biziura* compared to *Heteronetta*. The vertical lines at the top of the columns indicate two standard errors on either side of the mean.

tween the toes, though at the same time making terrestrial locomotion more clumsy and difficult. *Heteronetta* is intermediate in the development of this character between *Anas* and *Oxyura;* giving a clear indication of its phylogenetically and adaptively intermediate position between the Anatini and the advanced Oxyurini.

KNEE JOINT

Different groups of diving birds have convergently developed an elongated cnemial crest which is functionally similar but anatomically distinct in the different forms. In loons it is derived entirely from the tibiotarsus, in grebes from

TABLE 5

LENGTH OF THE INNER CNEMIAL CREST EXPRESSED AS A PERCENTAGE
OF THE LENGTH OF THE FEMUR

(The differences between the means of certain species were examined for statistical significance with the Student's t-test. The pairs tested are identified by superscripts in the Mean column, and the probability of the values belonging to the same population is given in the footnotes. Hence the difference between *Anas* and *Biziura* is highly significant, as is that between *Oxyura* and *Biziura*. The difference between *Anas* and *Heteronetta* was not significant. The same form of notation will be used throughout the tables, when several tests were made.)

Species	No. examined	Mean	Range
Anas platyrhynchos	10	10.43[a]	9.48–12.24
Heteronetta atricapilla	2	8.94	7.53–10.34
Oxyura jamaicensis	9	17.31[b]	14.79–19.06
Biziura lobata	7	14.04[a,b]	12.91–15.93

[a] $P < .001$.
[b] $P < .001$.

both the tibiotarsus and an enlarged patella, and in *Hesperornis* from the patella alone (Storer, 1960:28–29). A corresponding modification occurs in the stifftail ducks, involving both the tibiotarsus and the patella. In the different species there are different degrees of relative enlargement of both the inner cnemial crest of the tibiotarsus and of the patella, which together result in an extension of the axis of the tibiotarsus anterior to its proximal articular surface. This change allows for an anterior shift of the proximal areas of origin of certain muscles of the shank which become enlarged in diving ducks in conjunction with their importance in the movements of the tarsometatarsus in swimming. These are the M. gastrocnemius pars interna and M. peroneus longus, which extend the tarsus, and M. tibialis anterior, which flexes it. These will be described in detail below.

The inner cnemial crest was measured from its tip to its base at the articular surface of the tibiotarsus, and is expressed as a percentage of the length of the femur in table 5. Figure 13 shows the median side of the knee area in the four species under consideration. The inner cnemial crest is deep, hatchet-shaped, and relatively short in *Anas* and *Heteronetta*, but shallow and elongated in *Oxyura* and *Biziura*. In all three Oxyurini the patella is extensively ossified, while in *Anas* it is mostly ligamentous. The patella is slightly larger in *Heteronetta* than in *Oxyura* (table 6) resulting in an elongated tibial axis despite the short, *Anas*-like inner cnemial crest. *Biziura* has an enormous patella, extended so far that the

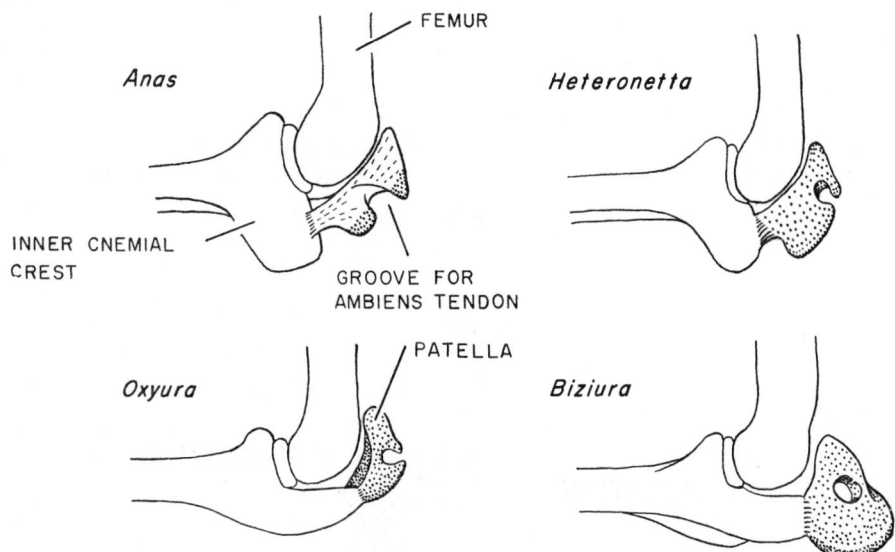

Fig. 13. Medial view of the knee area in *Anas platyrhynchos*, *Heteronetta atricapilla*, *Oxyura jamaicensis*, and *Biziura lobata*. The ossified region of the patella is stippled.

groove for the ambiens tendon has become entirely enclosed, a condition comparable to that in cormorants.

Total elongation anterior to the articular surface of the tibiotarsus is given by the sum of the lengths of the cnemial crest and patella. In *Heteronetta* this is 23.5 percent of femoral length; in *Oxyura* 30.95 percent; and in *Biziura* 43.73 percent. Thus this character shows increasing specialization for aquatic locomotion in the series of genera representing the evolutionary sequence within the Oxyurini.

MUSCLES OF THE HIND LIMB

The most important studies of the myology of the hind limb in waterfowl are those of Quennerstedt (1872), Stolpe (1932), Frantisek (1934), and Miller (1937). The first three provided descriptions of the muscles in ducks, while Miller and Stolpe gave both descriptions and functional analyses in geese and ducks respectively. The most thorough descriptive study of the Mallard is in an unpublished thesis by Philip Ruck (1949), who also made brief comparisons with the muscles of *Oxyura jamaicensis*, as well as numerous other species not considered in the

TABLE 6

LENGTH OF THE PATELLA EXPRESSED AS A PERCENTAGE OF THE LENGTH OF THE FEMUR[a]

Species	No. examined	Mean[b]	Range
Heteronetta atricapilla...............	2	14.51	12.44–16.58
Oxyura jamaicensis..................	4	13.64	12.59–15.13
Biziura lobata......................	6	29.70	27.66–32.53

[a] *Anas* is not included since the patella is largely unossified and a comparable measurement cannot be made accurately.
[b] The difference in the means of *Heteronetta* and *Oxyura* is not significant, but that between *Oxyura* and *Biziura* is ($P < .001$).

present study. Additional references to related studies are given by Humphrey and Clark (1964), who provide an excellent bibliography on all aspects of waterfowl anatomy. The muscle terminology used here is that of George and Berger (1966). Since there has been some confusion in the literature concerning the application of the terms flexion and extension to the movements of the different segments of the avian limb, the system used here is illustrated in figure 14.

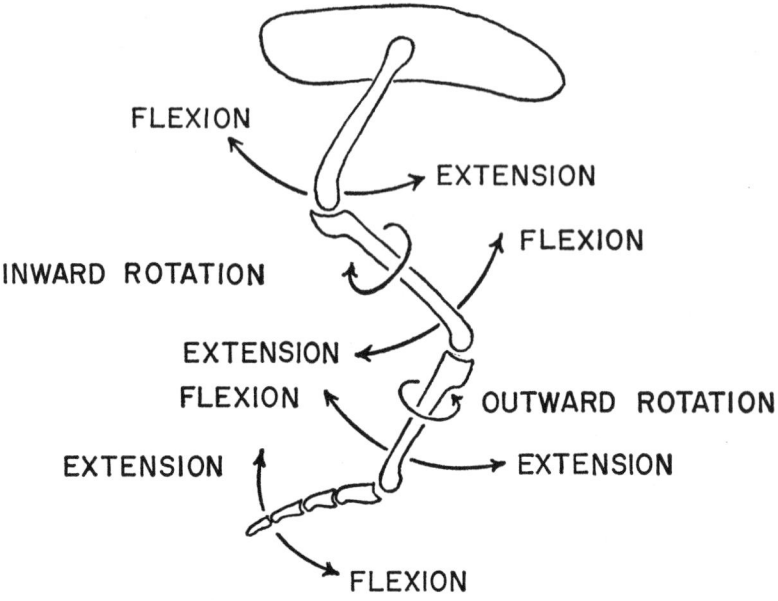

Fig. 14. Diagram illustrating the use of terminology referring to movements of the limb segments.

M. SARTORIUS (figs. 15, 19, 28, 30)

Structure.—This strap-shaped muscle forms the anterior margin of the thigh. It originates from the anterior crest of the ilium. The first third of the origin is fleshy, while the rest arises by an aponeurosis continuous with that of M. iliotibialis. The insertion is mainly fleshy on the antero-lateral surface of the knee joint, mostly on the patella but to a slight extent also on the internal cnemial crest.

Action.—Flexion of the femur, extension of the tibiotarsus, and adduction of the entire limb. It may also perform inward rotation of the tibiotarsus against the femur.

Comparison.—In the Oxyurini the origin is fleshy for its entire length. In *Biziura* the posterior one-third of the origin is overlain by the anterior edge of M. iliotibialis, while in the other species there is only a slight overlap.

M. ILIOTIBIALIS (figs. 15, 28)

Structure.—A thin, flat muscle on the lateral surface of the thigh posterior to M. sartorius and anterior to M. semitendinosus. The origin, from the iliac crest, is fleshy at the posterior end but aponeurotic for most of its length. This aponeurosis is triangular, with its apex lying distal to the underlying trochanter of the femur. The pre- and postacetabular portions are of about equal size. The postacetabular portion covers the origin of M. biceps femoris in contrast to *Branta* (Miller, 1937) in which it is exposed. At the origin some fibers fuse with M. biceps femoris posteriorly. Distally the muscle is divided into anterior and posterior fleshy insertions separated by an aponeurosis which is closely bound to the underlying M. femorotibialis externus et medius. This inserts on the patellar tendon and associated fascia on the lateral side of the knee joint.

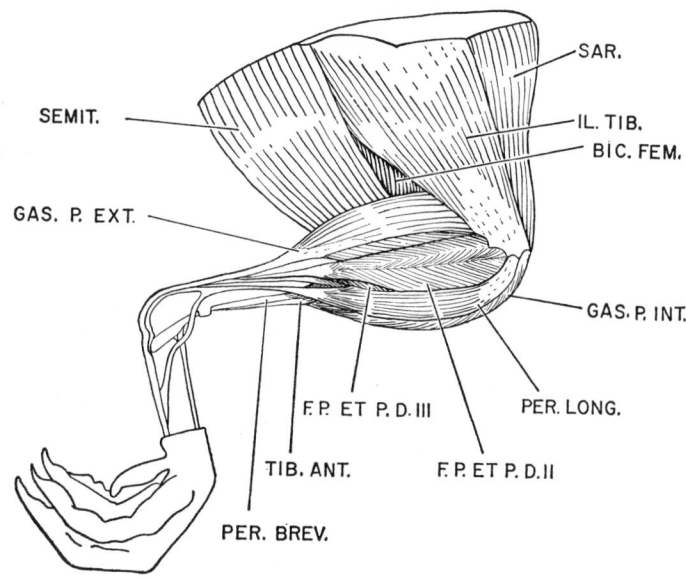

Fig. 15. First layer of muscles on the lateral surface of the hind limb of *Anas platyrhynchos*. The pubis, caudal skeleton, and M. piriformis pars caudofemoralis have been omitted for clarity.

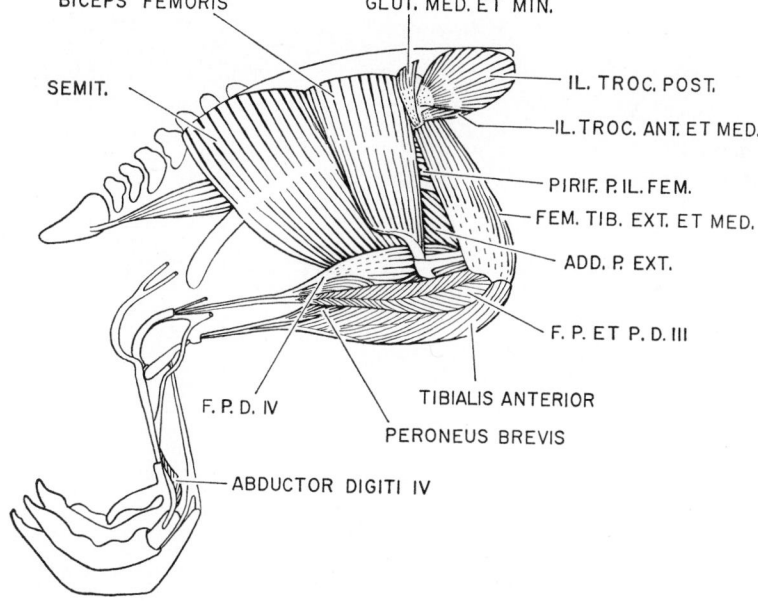

Fig. 16. Lateral view of a second layer of muscles in the hind limb of *Anas platyrhynchos*. The following muscles have been removed: sartorius, iliotibialis, gastrocnemius pars externa, gastrocnemius pars interna, peroneus longus, flexor perforans et perforatus digiti II.

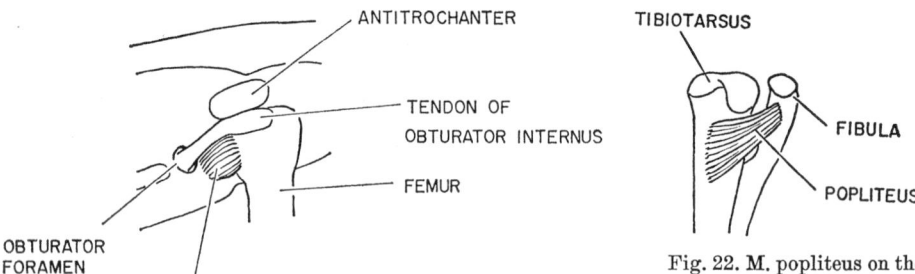

Fig. 21. M. obturator externus in *Anas platyrhynchos*.

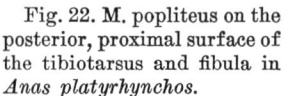

Fig. 22. M. popliteus on the posterior, proximal surface of the tibiotarsus and fibula in *Anas platyrhynchos*.

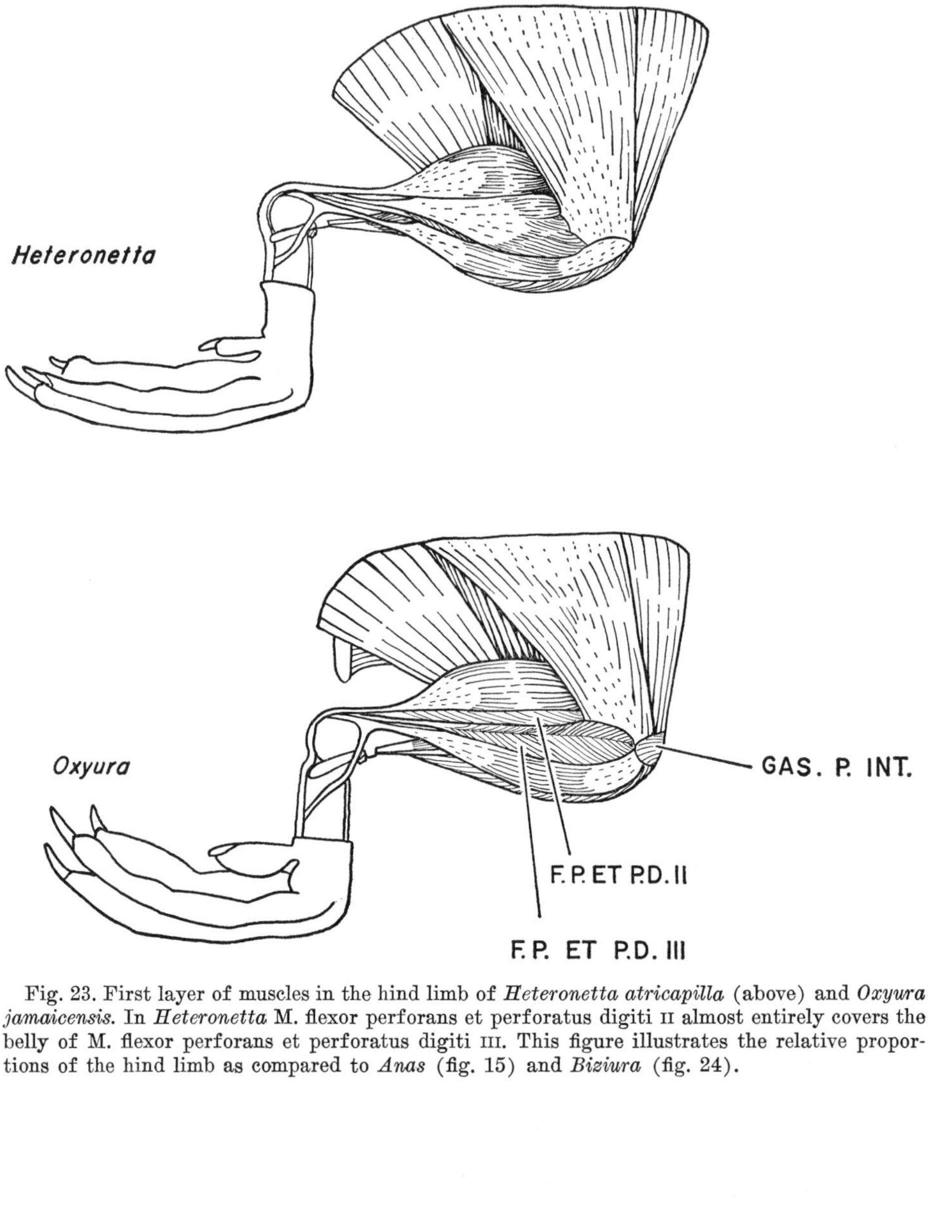

Fig. 23. First layer of muscles in the hind limb of *Heteronetta atricapilla* (above) and *Oxyura jamaicensis*. In *Heteronetta* M. flexor perforans et perforatus digiti II almost entirely covers the belly of M. flexor perforans et perforatus digiti III. This figure illustrates the relative proportions of the hind limb as compared to *Anas* (fig. 15) and *Biziura* (fig. 24).

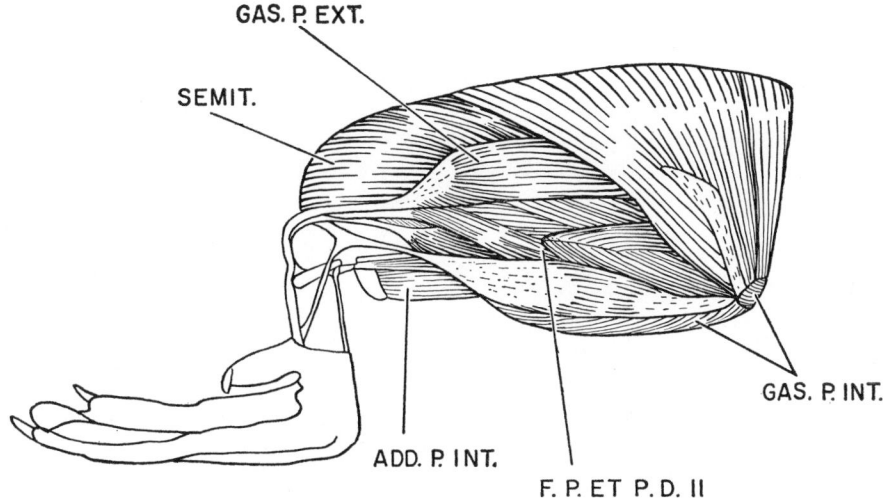

Biziura

Fig. 24. The first layer of muscles on the lateral surface of the hind limb of *Biziura lobata*. M. flexor perforans et perforatus digiti II is directly below M. gastrocnemius pars externa. The other muscles are as labeled in figure 15. This figure illustrates the difference in relative proportions of the hind limb of *Biziura* as compared to *Anas* (fig. 15), *Heteronetta* (fig. 23), and *Oxyura* (fig. 23).

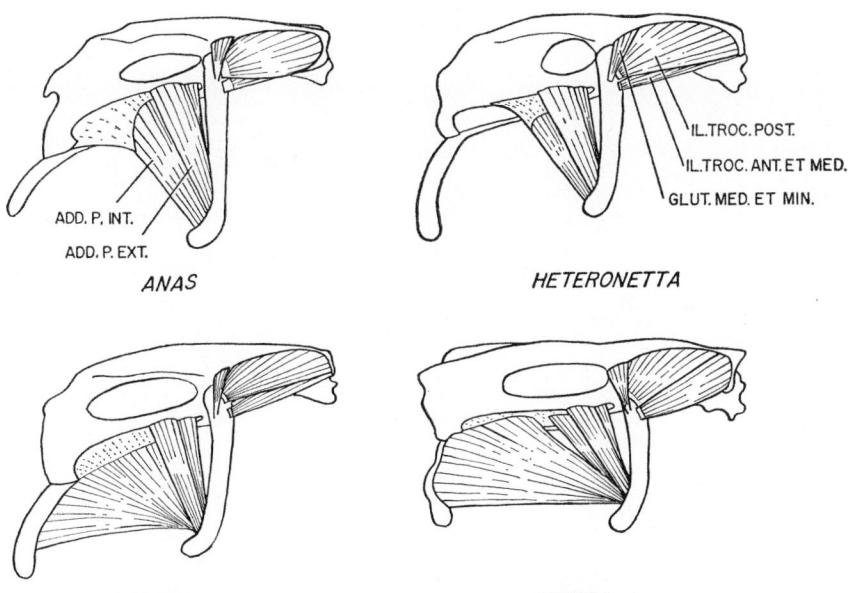

Fig. 25. Certain deep muscles of the thigh in *Anas platyrhynchos*, *Heteronetta atricapilla*, *Oxyura jamaicensis*, and *Biziura lobata*. In this sequence of genera there is a trend for the reduction in size of M. iliotrochantericus posterior and an increase in M. iliotrochantericus anterior et medius. In *Anas* and *Heteronetta* M. adductor longus et brevis pars interna arises by an aponeurosis from the ischium only, while in *Oxyura* and *Biziura* a fleshy origin has extended onto the pubis. The gap shown in the anterior part of this muscle in *Biziura* may have been due to damage in the specimen. Further details are given in the text.

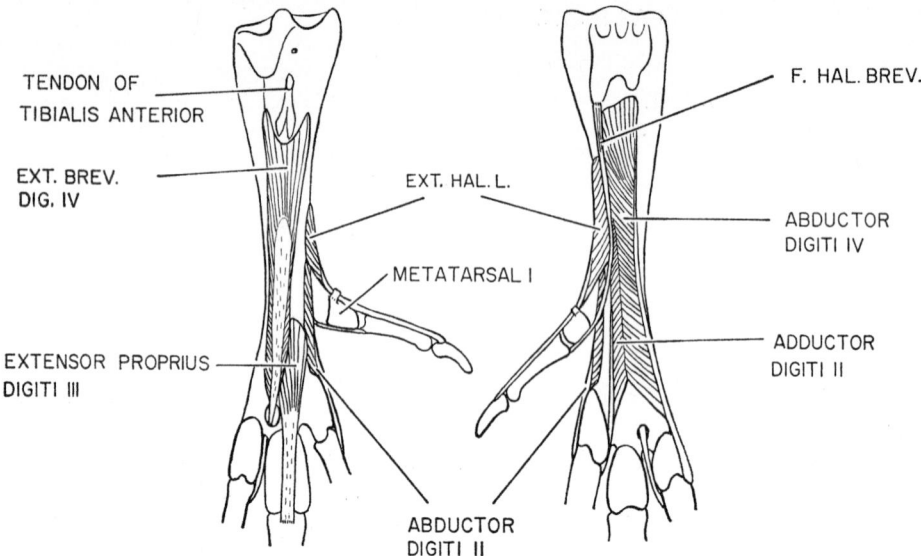

Fig. 26. Intrinsic muscles of the foot in *Anas platyrhynchos*. Left: anterior view of the tarsometatarsus. Right: posterior view of the tarsometatarsus.

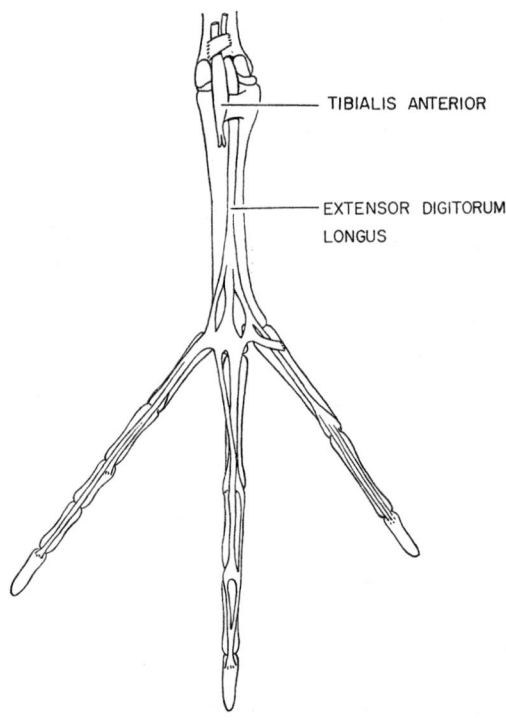

Fig. 27. Anterior view of the foot of *Anas platyrhynchos* showing the tendons of insertion of Mm. tibialis anterior and extensor digitorum longus.

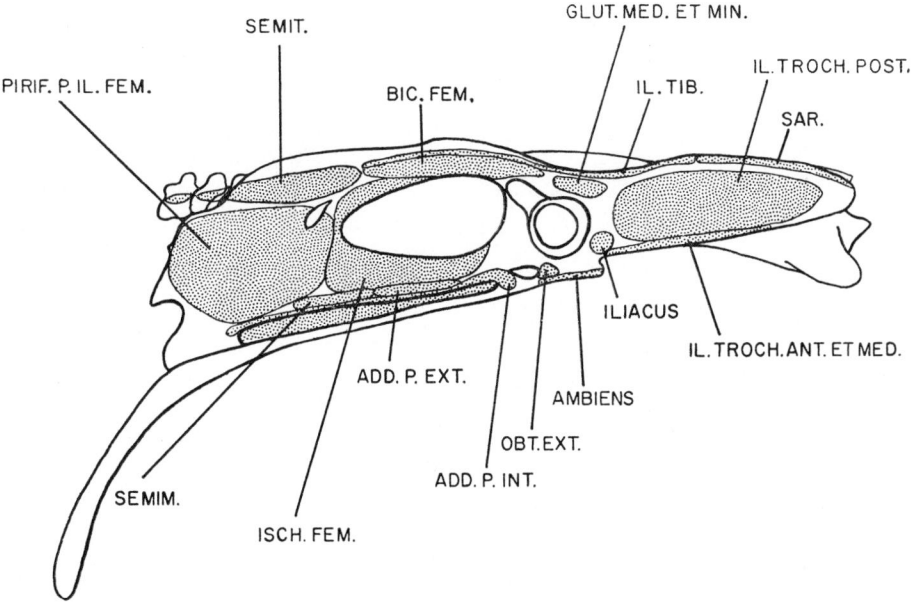

Fig. 28. Pelvic girdle of *Anas platyrhynchos* showing the areas of origin of the pelvic muscles.

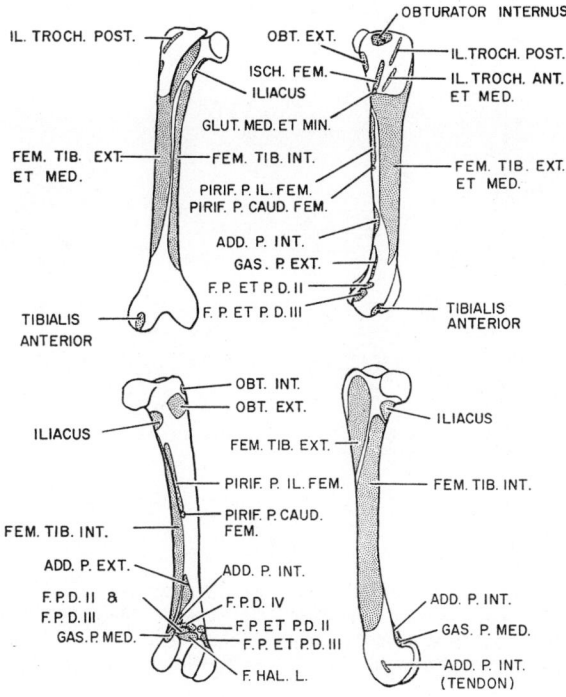

Fig. 29. Femur of *Anas platyrhynchos* showing the areas of attachments of muscles. The views are as follows: upper left, anterior; upper right, lateral; lower left, posterior; lower right, medial.

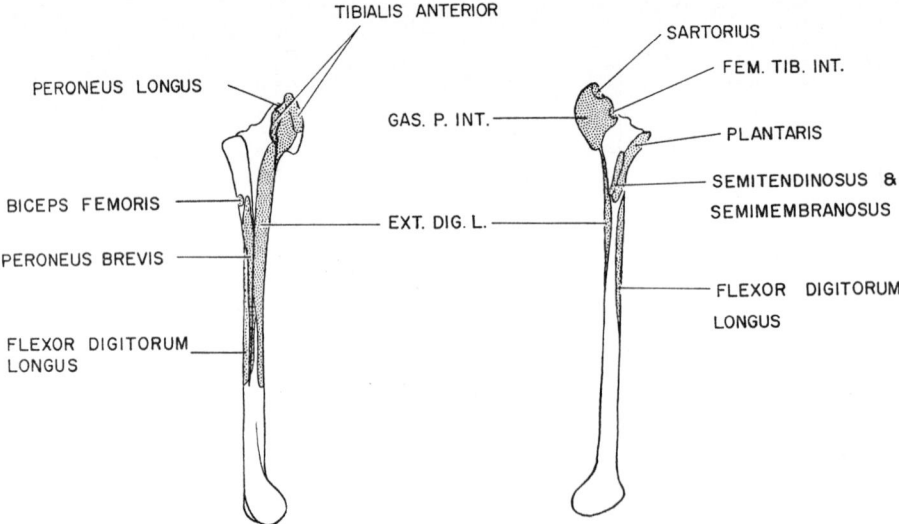

Fig. 30. Lateral (left) and medial views of the tibiotarsus and fibula of *Anas platyrhynchos* showing areas of muscle attachment.

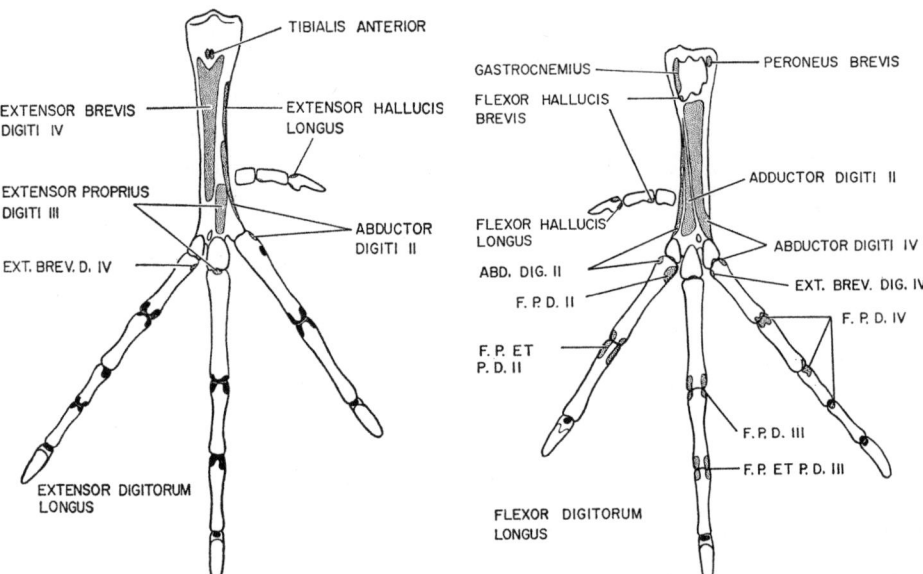

Fig. 31. Areas of attachment on the anterior surface of the foot of *Anas platyrhynchos*. The multiple insertions of M. extensor digitorum longus are shown in black.

Fig. 32. Areas of muscle attachment on the posterior side of the foot in *Anas platyrhynchos*. The points of insertion of M. flexor digitorium longus are indicated in black.

Action.—Primarily abduction of the limb, a function which it shares only with the very small M. gluteus medius et minimus. It may also extend the tibiotarsus slightly.

Comparison.—In *Oxyura* the fleshy fibers of the posterior margin of the origin do not extend medially to the midline, hence the posterior origin is entirely aponeurotic and there is no connection of this muscle to M. biceps femoris. In *Biziura* the aponeurosis of origin is reduced and extends distally only to the level of the trochanter. The posterior end of the origin is fleshy but only by a very few fascicles.

M. BICEPS FEMORIS (figs. 15, 16, 28, 30)

Structure.—The origin is fleshy from the posterior iliac crest just posterior to the acetabulum. The fibers converge distally to form a stout tendon, just proximal to which the muscle is attached by a few fibers to the surface of M. gastrocnemius pars externa. The tendon passes through the biceps loop to insert on the shaft of the fibula.

Action.—Flexion of the tibiotarsus and extension of the femur.

Comparison.—In *Biziura* the anterior margin of M. biceps femoris overlies most of the belly of M. gluteus medius et minimus.

M. GLUTEUS MEDIUS ET MINIMUS (figs. 16, 25, 28, 29)

Structure.—A small triangular muscle on the dorsolateral surface of the hip joint. The origin is fleshy from the posterior end of the anterior iliac crest. Insertion is by a flat tendon onto the proximo-lateral surface of the femur, just distal to the insertion of M. ischiofemoralis.

Action.—Abduction of the femur.

Comparison.—In *Heteronetta* the muscle is similar to that in *Anas*, with the fleshy portion extending distally only to the level of the antitrochanter, and giving rise to a long tendon of insertion which passes superficially to the insertion of M. ischiofemoralis. In *Oxyura* the muscle was fleshy for most of its length in three of four specimens, and similar to *Anas* in one. In *Biziura* the muscle is fleshy for most of its length, and inserts anterior to and at the same level as M. ischiofemoralis rather than distal to it as in the three other species.

M. ILIACUS (figs. 20, 28, 29)

Structure.—A small, strap-shaped muscle with a fleshy origin on the ventral edge of the preacetabular ilium, just posterior to the origin of M. iliotrochantericus anterior et medius, and just anterior to the acetabulum. It has a fleshy insertion on the postero-medial surface of the femur just distal to the neck.

Action.—Outward rotation of the femur.

Comparison.—Very small in *Biziura* (see table 15, below).

M. SEMITENDINOSUS (figs. 15, 16, 19, 24, 28, 30)

Structure.—A large, fan-shaped muscle on the lateral surface of the thigh posterior to M. iliotibialis. The origin is fleshy from the postero-dorsal edge of the ilium and the transverse processes of the first two free caudal vertebrae. The origin fuses slightly along its anterior border with M. biceps femoris. Distally the muscle forms a broad, flat tendon which fuses with M. semimembranosus. The common tendon thus formed inserts on the proximal end of the tibiotarsus between the bellies of M. gastrocnemius pars media and pars interna. There is no accessory semitendinosus.

Action.—Flexion of the tibiotarsus and extension of the femur.

Comparison.—In *Oxyura* and *Biziura* the origin is shifted somewhat posteriorly, especially in *Biziura*, where only half or less of the muscle arises from the ilium, the rest from caudal vertebrae. As a result of this shift, and probably compounded by the greater postacetabular length of the pelvis, the line of action of this muscle in *Oxyura* and especially in *Biziura* is shifted into a more horizontal plane than in *Anas* or *Heteronetta*.

M. SEMIMEMBRANOSUS (figs. 17, 18, 20, 28, 30)

Structure.—A strap-shaped muscle with a fleshy origin from the ventro-lateral margin of the ischium just posterior to the origin of M. adductor longus et brevis. Distally this muscle is fused with M. semitendinosus to form a common, broad, flat tendon which inserts onto the medial surface of the proximal end of the tibiotarsus.

Action.—Flexion of the tibiotarsus and extension of the femur.

Comparison.—Relatively narrower in *Oxyura* and *Biziura*.

M. AMBIENS (figs. 19, 28)

Structure.—This is the most superficial muscle on the medial surface of the thigh. The origin is by two fairly distinct heads. The anterior head arises by a tendon from the pectineal process of the pelvis, while the posterior head arises just posterior to it from the ventral edge of the pubis. The two heads are closely united for their entire length, and fuse distally, giving rise to a tendon which passes forward across the anterior surface of the knee in a groove in the patella, then continues posteriorly into the shank where it gives rise to lateral heads of Mm. flexor perforatus digiti II, III, and IV.

Action.—The main function is to supplement the contraction of the perforated flexors. There may also be adduction of the femur and extension of the shank.

M. FEMOROTIBIALIS EXTERNUS ET MEDIUS (figs. 16, 20, 29)

Structure.—A large muscle covering the lateral, anterior, and part of the medial surface of the femur. The two components are distinct at their origins and insertions, but tend to fuse for most of the length of their bellies. There is a fair amount of individual variation in the degree of fusion.

The origin is fleshy for most of the length of the shaft of the femur, including the lateral surface lateral to the anterior intermuscular line. Laterally the origin begins just distal to the insertion of M. ischiofemoralis. Anteriorly the origin extends across the intermuscular line and extends farther proximally medial to the trochanteric ridge.

Pars media inserts by a tendon on the medial, proximal edge of the patella. Pars externa inserts by a tendon on the anterior and lateral edge of the patella and patellar tendon, forming most of the latter; and fleshily on the dorsal surface of the patella. The tendon of M. ambiens enters the patella between the insertions of the two heads.

Action.—Extension of the tibiotarsus.

M. FEMOROTIBIALIS INTERNUS (figs. 20, 29, 30)

Structure.—There is a fleshy origin from the medial surface of the femur from the insertion of M. iliacus proximally to the median condyle distally. Insertion is by a tendon to the dorso-medial edge of the inner cnemial crest of the tibiotarsus.

Action.—Rotation of the tibiotarsus inward. Because of its insertion at the joint it appears by manipulation to produce slight extension of the tibiotarsus when the latter is extended, and flexion when it is flexed. Its strength in either action must be very slight.

Comparison.—In *Anas* and *Heteronetta* the belly is relatively narrow, its posterior border relatively straight and barely overlapping the insertion of M. adductor longus et brevis. In both cases also a tendinous sheet overlies the distal portion of the muscle and extends to the posterior margin. In *Oxyura* and *Biziura* the belly is relatively wider and overlaps the insertion of M. adductor longus et brevis to a greater extent. In these genera the posterior margin of the muscle is fleshy rather than overlain by a tendinous sheet.

M. PIRIFORMIS PARS ILIOFEMORALIS (figs. 16, 17, 28, 29)

Structure.—The origin is fleshy from the postero-lateral surface of the pelvis in the area of fusion of the ilium and ischium, ventral to the origin of M. semitendinosus and posterior to the origin of M. ischiofemoralis. Insertion is by a broad tendon onto the posterior surface of the femur distal to the insertion of M. ischiofemoralis.

Action.—Extension of the femur.

Comparison.—In *Oxyura* and *Biziura* the origin is located farther posteriorly due to the greater postacetabular length of the pelvis. As a result the line of action of the muscle is slightly more horizontally aligned. In *Biziura* the tendon of insertion is quite narrow compared to the other species (3.9 mm compared to 12–15 mm).

M. Piriformis pars Caudofemoralis

This muscle is described in the section on myology of the tail, above.

M. Ischiofemoralis (figs. 17, 18, 28, 29)

Structure.—A broad, fan-shaped muscle arising from the lateral surface of the ischium posterior to the acetabulum, and from the membrane of the ilio-ischiadic fenestra. It lies deep to M. piriformis pars iliofemoralis. The anterior two-thirds of the belly is covered with a dense fascia which contributes to the tendon of insertion on the lateral surface of the femur posterior to, and midway between the insertions of Mm. iliotrochantericus posterior and iliotrochantericus anterior et medius.

Action.—Outward rotation, and possibly slight extension of the femur.

Comparison.—The area of origin is shifted somewhat posteriorly in *Oxyura* and *Biziura*.

M. Adductor Longus et Brevis (figs. 16, 17, 18, 19, 20, 25, 29)

Structure.—This is a broad, flat muscle composed of a larger medial portion (pars interna) and a smaller lateral portion (pars externa). Pars interna originates from the ischium and pubis just below and posterior to the obturator foramen (fleshy) and by an aponeurosis from the latero-ventral edge of the ischium and ischiopubic membrane. Pars externa takes its fleshy origin from the ventral anterior edge of the ischium posterior to the anterior part of pars interna.

Pars interna inserts by a tendon on the medial surface of the internal condyle of the femur, and, partly in common with a fused portion of M. gastrocnemius pars media on the posterior edge of the condyle just proximal to the rest, this portion being fleshy. Pars externa has a mixed insertion on the posterior surface of the femur proximal to the insertion of pars interna. Pars interna and pars externa tend to fuse along the anterior edge of pars externa so that it is difficult to separate them; for this reason the entire muscle is weighed as a unit. The degree of fusion varies with the individual.

Action.—Because the insertion is on the posterior face of the femur, the primary action would appear to be extension of the femur rather than adduction, despite the name of the muscle. Some adduction might occur when the femur is strongly abducted.

Comparison.—The condition in *Heteronetta* is essentially as described for *Anas*, but in *Oxyura* and *Biziura* the muscle shows a greater degree of modification than any other muscle of the hind limb. Pars externa remains small, but pars interna has extended its origin far posteriorly along the pre- and postischiadic pubis; this origin furthermore is largely fleshy from the pubis and ischiopubic membrane rather than aponeurotic (fig. 25). As a result, the line of action of the muscle is rotated into a much more horizontal position; the functional significance of this change is discussed below (p. 40).

M. Obturator Internus (figs. 18, 19, 21, 29)

Structure.—A bipennate muscle whose fibers converge upon a central tendon three-fourths as long as the belly. This muscle occupies the ventral part of the ischiopubic fenestra, originating from the ventro-medial rim of the ischium, the dorso-medial rim of the pubis, and the intervening ischiopubic membrane. The tendon passes through the obturator foramen, accompanied by a few fibers, and inserts on the dorsal edge of the trochanter of the femur proximal to the insertion of M. ischiofemoralis.

Action.—Outward rotation of the femur.

Comparison.—The posterior margin of the muscle has a rounded outline in *Anas*, but is sharply squared off in the other genera.

M. Obturator Externus (figs. 18, 21, 28, 29)

Structure.—This very small muscle arises from the postero-ventral margin of the acetabulum just anterior to the obturator foramen. It passes laterally and inserts on a small, roughly rectangular rugosity on the posterior, proximal surface of the femur. The origin and insertion are both fleshy.

Action.—Supplements the action of M. obturator internus.

M. GASTROCNEMIUS (figs. 15, 19, 20, 23, 24, 29, 30, 32)

Structure.—A very large muscle superficially located on the posterior surface of the shank. There are three parts:

Pars externa: This occupies the posterio-lateral surface of the shank. It originates from a bony ridge on the postero-lateral surface of the distal end of the femur, via a broad tendon. It gives rise to a tendon which fuses near the tibial cartilage with the tendon of pars interna.

Pars interna: The largest of the three bellies, this is the most superficial muscle on the medial surface of the shank. There are two heads. The posterior head arises from the inner cnemial crest of the tibiotarsus, the anterior one from the patella. The tendon of pars media enters the belly near its distal end.

Pars media: This is the smallest of the three parts of M. gastrocnemius. It lies just lateral to pars interna, but separated from it by the distal end of Mm. semitendinosus and semimembranosus. The origin is fleshy, in common with the insertion of M. adductor longus et brevis, from the posterior edge of the median condyle of the femur. The tendon of insertion fuses with that of pars interna.

The common tendon of pars media and pars interna passes distally and fuses with that of pars externa near the distal end of the tibiotarsus, to form a common Tendon of Achilles. This passes over the tibial cartilage, at which point it is widened and finally inserts on the hypotarsus. It is also closely fused here with a tendinous sheet which covers the flexor tendons.

Action.—The major function is powerful extension of the tarsometatarsus. Flexion of the tibiotarsus upon the femur may also be accomplished because of the origins of pars externa and pars media from the femur. This may counteract the tendency for the power stroke in swimming, tarsal extension, to cause extension of the tibiotarsus.

Comparison.—In *Oxyura* and *Biziura* the belly of pars externa is relatively longer than in *Anas* or *Heteronetta*. In *Oxyura* and *Biziura* the proximal end of the anterior head of pars interna arises on the lateral surface of the knee, passing around the knee to the medial side where the belly lies. In *Heteronetta* pars media was absent in both legs of the one specimen dissected.

M. ILIOTROCHANTERICUS POSTERIOR (figs. 16, 25, 28, 29)

Structure.—A stout, fan-shaped muscle occupying the preacetabular ilium. The origin is fleshy from the anterior iliac fossa. The fibers converge to form a flat tendon which inserts on the proximal end of the trochanter of the femur at about the level of the acetabulum.

Action.—Inward rotation and possibly a slight flexion of the femur.

Comparison.—In *Heteronetta* and *Oxyura* the outer margin is not as rounded as in *Anas* because of the greater anterior extension of M. iliotrochantericus anterior et medius. This trend is seen to an even greater extent in *Biziura* (fig. 25).

M. ILIOTROCHANTERICUS ANTERIOR ET MEDIUS (figs. 16, 17, 19, 25, 28, 29)

Structure.—This muscle is formed by a fusion of M. iliotrochantericus anterior and M. iliotrochantericus medius, which are separate in some groups of birds. The mixed origin arises from the ventral edge of the anterior iliac fossa, and is mostly covered by M. iliotrochantericus posterior. The fibers converge to insert via a flat tendon on the lateral surface of the femur just distal to the insertion of M. iliotrochantericus posterior.

Action.—Flexion and inward rotation of the femur.

Comparison.—In *Oxyura* and *Biziura* the origin is broader and extends farther dorsally at its anterior end. In some specimens of *Oxyura* the muscle is fairly distinctly divided into two parts, presumably representing the original pair of muscles.

M. FLEXOR PERFORANS ET PERFORATUS DIGITI II (figs. 15, 17, 23, 29, 32)

Structure.—A multipennate muscle on the lateral surface of the shank posterior to M. peroneus longus and anterior to M. gastrocnemius pars externa. There are two parts. The medial head arises from the external condyle of the femur and the lateral head arises from the lateral surface of the knee joint, overlying the origin of M. flexor perforans et perforatus digiti III. It forms a tendon which passes over the tibial cartilage and forward along the ventral surface of the tarsometatarsus and digit II. At the end of the first phalanx it bifurcates and inserts mainly on the proximal end of the second phalanx and slightly also on the distal end of the first. It is perforated here by the tendon of M. flexor digitorum longus.

Action.—Flexion of the distal part of digit II if the proximal phalanx is fixed, if not, flexion of the entire digit II; also abduction of digit II. Extension of the tarsometatarsus if digit II is fixed.

Comparison.—In *Heteronetta* the muscle consists of two heads as in *Anas*, but the anterior end of the lateral head does not extend quite so far onto the knee joint. In *Oxyura* only the medial head is present. In *Biziura* a small part of the lateral head remains, arising only from the fascial covering of M. flexor perforans et perforatus digit III (fig. 24).

M. FLEXOR PERFORANS ET PERFORATUS DIGIT III (figs. 16, 17, 23, 29, 32)

Structure.—A bipennate muscle lying on the lateral surface of the shank deep to M. flexor perforans et perforatus digiti II. It arises fleshily from the lateral surface of the patellar tendon, by a tendon from the lateral surface of the external condyle of the femur, and by a fleshy origin from the lateral surface of the fibula distal to the insertion of the biceps tendon. It forms a tendon which passes over the tibial cartilage and ventrally along the tarsometatarsus and third digit. It perforates the tendon of M. flexor perforatus digiti III and then bifurcates to insert on the distal end of phalanx II and the proximal end of phalanx III of digit III. Here it is perforated by the tendon of M. flexor digitorum longus.

Action.—Flexion of digit III; extension of the tarsometatarsus when digit III is fixed in position by its extensors.

M. PERONEUS LONGUS (figs. 15, 30)

Structure.—This muscle is superficially located on the antero-lateral surface of the shank, posterior to M. gastrocnemius pars interna and anterior to M. flexor perforans et perforatus digiti II. It arises by an aponeurosis from the head of the tibiotarsus, the inner and outer cnemial crests, and the patella. It gives rise to a flat tendon which passes along the lateral surface of the tibiotarsus, at the distal end of which it bifurcates. One wide, short branch inserts on the proximo-lateral surface of the tibial cartilage. The main tendon continues across the lateral surface of the intertarsal joint and passes to the posterior side of the tarsometatarsus where it fuses with the tendon of M. flexor perforatus digiti III.

Action.—Extends the tarsometatarsus and strengthens the action of M. flexor perforatus digiti III.

Comparison.—In *Oxyura* and *Biziura* the belly is relatively longer than in the other genera, reaching nearly to the end of the tibiotarsus.

M. TIBIALIS ANTERIOR (figs. 15, 16, 20, 27, 29, 30, 31)

Structure.—This muscle lies on the anterior surface of the tibiotarsus superficial to M. extensor digitorum longus and deep to M. peroneus longus. The tibial head arises from the inner and outer cnemial crests and the lateral surface of the patella between them. This origin is mainly fleshy. The femoral head arises by a short tendon from the anterior surface of the lateral condyle of the femur. The tendon of insertion passes distally in the tendinal groove of the tibiotarsus, and just proximal to the condyles of the tibiotarsus it passes under a stout tendinous loop, the transverse ligament. It then becomes somewhat wider as it crosses the intertarsal joint and bifurcates just before inserting on a tubercle on the proximal end of the anterior face of the tarsometatarsus.

Action.—Flexion of the tarsometatarsus. The lateral head may also flex the tibiotarsus due to its origin on the femur.

Action.—Since the fibula is relatively fixed in position, it is probable that the function of this muscle is mainly in fixation rather than movement. Perhaps it serves to maintain the head of the fibula in contact with the fibular groove of the femur during walking, as suggested by Fisher (1946) in his study of Cathartid vultures.

M. EXTENSOR HALLUCIS LONGUS (figs. 20, 26, 31)

Structure.—A small, elongate, fan-shaped muscle which arises on the postero-medial surface of the tarsometatarsus, beside the ventral bundle of flexor tendons and proximal to M. abductor digiti II. The tendon passes obliquely forward across the posterior part of the first metatarsal after passing beneath a very delicate sling which arises from the metatarsal. A short branch attaches to this bone, but the main tendon continues across the upper surface of the first phalanx and inserts at the base of the ungual phalanx.

Action.—Extension of the hallux.

Comparison.—The muscle is very small in *Biziura*.

M. ABDUCTOR DIGITI II (figs. 20, 26, 31, 32)

Structure.—A small, fan-shaped muscle arising on the medial surface of the tarsometatarsus distal to M. extensor hallucis longus. It inserts by a tendon onto the medial side of the proximal end of the first phalanx of digit II.

Action.—Abducts digit II.

M. ADDUCTOR DIGITI II (figs. 26, 32)

Structure.—A small, pennate muscle with a fleshy origin from the ventral surface of the tarsometatarsus medial to the origin of M. abductor digiti IV. A fine tendon arises from the medial surface of the muscle, passes between the trochleae of digits II and III, and inserts on the medial proximal surface of the first phalanx of digit II.

Action.—Adduction of digit II.

M. EXTENSOR PROPRIUS DIGITI III (figs. 26, 31)

Structure.—A very small, fan-shaped muscle arising from the dorsal face of the distal one-fourth of the tarsometatarsus, just proximal to the trochlea of digit III and medial to the belly of M. extensor brevis digiti IV. It gives rise to a broad, flat tendon which passes over the trochlea and inserts on the base of the first phalanx of digit III.

Action.—Extends digit III.

M. FLEXOR HALLUCIS BREVIS (figs. 26, 32)

Structure.—A tiny, extremely slender muscle arising from the medial distal margin of the hypotarsus and giving rise to a fine tendon which inserts on the ventral side of the base of the first phalanx of the hallux.

Action.—Flexes the hallux.

M. EXTENSOR BREVIS DIGITI IV (figs. 26, 31)

Structure.—This long, narrow muscle arises fleshily from the anterior surface of the tarsometatarsus beginning just distal to the insertion of M. tibialis anterior. The tendon passes through the distal foramen of the tarsometatarsus to emerge between the trochleae of digits III and IV and inserts on the medial surface of the base of phalanx I of digit IV.

Action.—Extension, and possibly adduction of digit IV.

M. ABDUCTOR DIGITI IV (figs. 16, 26, 32)

Structure.—This muscle has a fleshy origin from the posterior surface of the tarsometatarsus distal to the hypotarsus, and spreading onto the distal one-fifth of the lateral surface of the tarsometatarsus proximal to the trochlea of digit IV. It inserts by a tendon onto the latero-ventral corner of the proximal end of phalanx I of digit IV.

Action.—Abduction of digit IV.

M. LUMBRICALIS (fig. 18)

Structure.—This is the smallest muscle of the hind limb, being only about 2.5 mm long when stretched and less than 1 mm in diameter. It arises on the inner surface of the tendon of M. flexor digitorum longus just proximal to the trifurcation of this tendon. There are two strap-shaped bellies, one inserting on the joint pully of the third digit, the other on that of the fourth digit. The former is the larger of the pair.

Action.—It is difficult to imagine what role this muscle might have in view of its small size. Possibly it functions in some way in delicate adjustments of the tendon from which it arises.

Comparison.—The muscle is even smaller in *Heteronetta*, *Oxyura*, and *Biziura* than it is in *Anas*.

DISCUSSION

The most distinct myological difference found among the four species dissected was the absence of M. gastrocnemius pars media in *Heteronetta*. It was not present in either leg of a single specimen, but other specimens should be examined to verify that this is a species characteristic rather than an individual anomaly. In any event, M. gastrocnemius pars media is probably of little functional significance because of its small size in the other genera and its lack of an independent insertion.

A thorough understanding of the functional differences of the muscular system in different species would require investigation of many factors such as blood supply, biochemistry, fiber type and arrangement, and the sequences and strengths of muscular actions in the living state. These will not be considered here. Nevertheless much can be learned from certain gross morphological features, including the relative size of the muscles (measured as dry weight as explained above in the Materials and Methods section) and their lever actions. The mechanical advantage of a muscle is equal to the force arm (distance from the fulcrum to the insertion of the muscle) divided by the resistance arm (distance from the fulcrum to the end of the element). A muscle which is inserted near the fulcrum will move the end of the element through a longer arc than one which is inserted farther from the fulcrum. If the speed of contraction is the same in both cases, the former element will thus move more rapidly in order to cover the greater distance in the same time (Alexander, 1968:13). Hence a larger mechanical advantage indicates adaptation for strength over speed, while a smaller value indicates adaptation for speed at the expense of strength. In the following analysis the major muscle groups will be considered according to the movements that they produce in the limb and correlated with the uses of the limb in locomotion as described above in the Aquatic Locomotion section.

MOVEMENTS OF THE FEMUR

The only important muscles which perform flexion of the femur without a direct action on other segments of the leg are the iliotrochantericus posterior and iliotrochantericus anterior et medius. Because of their insertion on the lateral face of the femur they may also cause inward rotation of the femur about its long axis. In both actions they are opposed by the obturator muscles and the ischiofemoralis. It is probable that a major function of all these muscles is stabilization of the hip joint. The iliotrochanterici are progressively diminished in relative weight as follows (mean values); *Anas,* 5.49 percent of total leg muscle weight;

Heteronetta, 5.16 percent; *Oxyura,* 4.70 percent; *Biziura,* 3.71 percent. The size of the two muscles relative to each other also change; in *Anas* the iliotrochantericus anterior et medius weighs 31.6 percent as much as the iliotrochantericus posterior; in *Heteronetta,* 48.7 percent; in *Oxyura,* 76.9 percent; and in *Biziura,* 62.6 percent. Table 15 (below) shows that this change is due mainly to a progressive reduction in the size of M. iliotrochantericus posterior while the iliotrochantericus anterior et medius does not diminish, and may even increase in weight.

The iliotrochantericus posterior is inserted at about the level of the head of the femur, hence its function must lie mainly in femoral rotation rather than flexion.

TABLE 7

MECHANICAL ADVANTAGE OF M. ILIOTROCHANTERICUS ANTERIOR ET MEDIUS

Species	No. examined	Mean	Range
Anas platyrhynchos	3	.202[a]	.194–.216
Heteronetta atricapilla	1	.196	....
Oxyura jamaicensis	3	.236[a,b]	.234–.237
Biziura lobata	3	.316[b]	.303–.331

[a] $P < .01$.
[b] $P < .02$.

The iliotrochantericus anterior et medius is inserted below the joint however, and must therefore contribute much more to flexion of the femur. Furthermore, there is a progressive shift in the insertion of this muscle so as to increase its mechanical advantage (table 7).

The amount of strength needed for fixation of the hip joint is probably much smaller in purely aquatic forms since the weight of the body is not supported here as it is in terrestrial locomotion, and this may account for the reduction in size of both the iliotrochantericus posterior and its major antagonist, M. obturator internus (table 15). M. ischiofemoralis is also antagonistic to M. iliotrochanericus posterior, and is considerably reduced in *Biziura* (table 15), yet in *Oxyura* it is significantly enlarged. The reason for this is not clear. Flexion of the femur is important in the back-and-forth, piston-like movement of the shank which augments the actions of the foot in swimming (see above, Summary of Locomotor Habits section) and this is apparently the reason that the iliotrochantericus anterior et medius has not undergone a similar reduction in relative size.

Direct extension of the femur is produced by Mm. piriformis pars iliofemoralis and adductor longus et brevis. There is some variation in the weights of these muscles relative to each other (table 15) but together they comprise a similar percentage of the total muscle weight in *Anas* (9.61 percent), *Heteronetta* (9.65 percent), and *Oxyura* (10.03 percent), but are much reduced in *Biziura* (6.42 percent). Extension of the femur is of great importance in walking on land, but of less importance in swimming (see above). However these muscles may have a very important function in opposing the tendency of the femur to be flexed when the tarsometatarsus is extended during swimming. The area of origin of the adductor is shifted posteriorly and ventrally in *Oxyura* and *Biziura* compared to the other species (fig. 25), and this has the effect of rotating the line of action of the muscle into a much more horizontal position, where it is less effective in extending the femur beyond the vertical, but much more efficient in opposing its flexion beyond that position. The insertion of the piriformis is also shifted distally in *Biziura*, thus increasing its mechanical advantage (table 8) and adding to the force with which it can act on the femur. Both of these muscles are in addition more effective in *Oxyura* and *Biziura* in extending the femur through a short distance in the action by which it imparts a piston-like movement to the tibiotarsus in strong swimming.

TABLE 8

MECHANICAL ADVANTAGE OF M. PIRIFORMIS PARS ILIOFEMORALIS

Species	No. examined	Mean	Range
Anas platyrhynchos	3	.379	.362–.399
Heteronetta atricapilla	1	.356	
Oxyura jamaicensis	3	.374	.365–.379
Biziura lobata	1	461	

MOVEMENTS OF THE TIBIOTARSUS

The primary extensors of the tibiotarsus are the sartorius and femorotibialis externa et medius. The sartorius produces both tibial extension and femoral flexion, unless the movement of the femur is opposed by its extensors. Because extension of the tibiotarsus is slight during swimming the sartorius appears to be of greater importance in terrestrial than in aquatic locomotion (Miller, 1937:26) and is found to be reduced in size in *Oxyura* compared to *Anas* and *Heteronetta*, and even further reduced in *Biziura* (table 15).

M. femorotibialis externus et medius is the only muscle causing extension of the tibiotarsus without involving other segments of the limb. Possibly it is aided slightly by M. femorotibialis internus, but this is probably of minor importance. The femorotibialis externus et medius is important in extension of the lower limb against the ground in walking, but, like the sartorius, its actions are minimal during swimming, and it also is reduced in the more aquatic species (table 15). Together the sartorius and femorotibialis externus et medius constitute the following percentages of limb muscle weight: *Anas*, 14.20 percent; *Heteronetta*, 12.46 percent; *Oxyura*, 9.05 percent; and *Biziura*, 8.58 percent.

Flexion of the tibiotarsus is produced by the biceps femoris and the semitendinosus and semimembranosus. The latter two muscles act together through a common insertion, and the relatively small semimembranosus is apparently just an auxiliary to the very large semitendinosus. These tibial flexors also produce femoral extension unless this is opposed by the femoral flexors. Powerful tibial flexion is important in walking, but in swimming the tibia moves only slightly and the main function of these muscles is apparently to oppose tibial extension caused by the extension of the tarsus in the main swimming movement.

Miller (1937) found that the biceps is enlarged in more aquatic species of geese, but the opposite is true in the species studied here (table 15). This is offset to some degree, however, by an increase in the mechanical advantage of the biceps

TABLE 9
MECHANICAL ADVANTAGE OF M. BICEPS FEMORIS

Species	No. examined	Mean	Range
Anas platyrhynchos	4	.240[a,b]	.233–.247
Heteronetta atricapilla	3	.229[a]	.218–.239
Oxyura jamaicensis	6	.253[b,c]	.247–.264
Biziura lobata	7	.271[c]	.266–.280

[a] $P < .20$.
[b] $P < .02$.
[c] $P < .001$.

TABLE 10
MECHANICAL ADVANTAGE OF M. SEMITENDINOSUS

Species	No. examined	Mean	Range
Anas platyrhynchos	3	.142	.137–.149
Heteronetta atricapilla	3	.147	.141–.153
Oxyura jamaicensis	3	.160	.117–.197
Biziura lobata	2	.152	.146–.157

(table 9) in *Oxyura* compared to *Anas* and *Heteronetta*, and in *Biziura* compared to *Oxyura*. The biceps is 5 percent more effective in producing force at the expense of speed per unit of contraction in *Oxyura* compared to *Anas*, while *Biziura* is 7 percent more effective than *Oxyura* and 13 percent more effective than *Anas*. *Heteronetta* does not differ significantly from *Anas* in mechanical advantage of the biceps.

The semitendinosus shows a tendency for increase in weight except in *Oxyura* (table 15), and also shows an increase in mechanical advantage, especially in *Oxyura* (table 10), which may compensate somewhat for the lack of increase in size. These differences however, are not statistically significant, at least in the small sample tested. This muscle is also shifted posteriorly in *Oxyura* and *Biziura* owing to the elongation of the postacetabular pelvis (see above, The Pelvic Girdle). This brings its line of action into a more horizontal plane which makes it less effective in opposing tibial extension, but more effective in opposing femoral flexion in a manner similar to the condition in the adductor longus et brevis in these species. The sum of the percent weights of Mm. biceps femoris, semitendinosus, and semimembranosus are as follows: *Anas*, 18.39 percent; *Heteronetta*, 20.83 percent; *Oxyura*, 15.04 percent; and *Biziura*, 16.79 percent. Thus it seems that the more limited use of this muscle group in *Oxyura* and *Biziura* requires less bulk. The difference between these two is not great, and perhaps the larger biceps in *Oxyura* compensates for the smaller size of its semitendinosus compared to *Biziura*.

MOVEMENTS OF THE FOOT

The primary power stroke in swimming and diving is extension of the tarsometatarsus, which is the function of M. gastrocnemius. This muscle is enlarged in the more aquatic species (table 14) with the greatest increase occurring in pars interna (table 15), which is correlated with the anterior extension of its area of origin on the inner cnemial crest and patella (see the description of the knee joint, above, in the Osteology of the Hind Limb section). In *Oxyura* and *Biziura* the medial portion of this muscle extends anteriorly around the knee to originate on the lateral surface (figs. 23 and 24); presumably this contributes to the increased bulk of the muscle noted above. This character is absent in *Anas* and *Heteronetta* (figs. 15 and 23).

M. gastrocnemius inserts on the hypotarsus of the tarsometatarsus, hence in its action of extending the latter, the length of the tarsometatarsus is the resistance

TABLE 11
Mechanical Advantage of M. Gastrocnemius

Species	No. examined	Mean	Range
Anas platyrhynchos	10	.079[a]	.073–.088
Heteronetta atricapilla	2	.091[a,b]	.090–.092
Oxyura jamaicensis	9	.108[b,c]	.099–.141
Biziura lobata	7	.135[c]	.119–.145

[a] $P < .001$.
[b] $P < .01$.
[c] $P < .001$.

arm, while the height of the hypotarsus represents the force arm. This was measured as the height of the inner (highest) calcaneal ridge to the foramen at its base. Table 11 shows that the sequence of genera considered to exhibit increasing aquatic specialization shows a progressive modification toward increased mechanical advantage in M. gastrocnemius. From this one may calculate that the gastrocnemius in *Heteronetta* produces 15 percent more force per unit of contraction than in *Anas*, but is correspondingly slower in action. In *Oxyura* and *Biziura* the values are 37 percent and 71 percent respectively. Thus it appears that force rather than speed of contraction in tarsal extension is advantageous in aquatic locomotion compared to walking. Miller (1937) found the same to be true in geese, although the range of difference was much smaller than in the species studied here. This change in lever action augments the increased size of M. gastrocnemius in producing a progressively greater force of contraction in diving species.

In addition to M. gastrocnemius, the flexor muscles of the foot may also act in tarsal extension. Most of these show an increase in relative weight in the series of genera *Anas, Heteronetta, Oxyura, Biziura* (tables 14 and 15). This increase not only parallels but exceeds the corresponding increase in size of the gastrocnemius (table 16).

M. flexor perforans et perforatus digiti II includes two distinct heads (femoral and patellar) in *Anas* and *Heteronetta*, but in *Oxyura* only the femoral head is present (fig. 23), while in *Biziura* (fig. 24) the posterior part of the patellar head is still present. Loss or reduction of the patellar part in *Oxyura* and *Biziura* is however, more than compensated for by an increase in the relative weight of the remaining portion (table 15).

In swimming, the digital flexors probably act in two ways. They aid in exten-

TABLE 12

Dry Weight of M. Tibialis Anterior Expressed as a Percent of the
Dry Weight of the Total Leg Musculature

Species	No. examined	Mean	Range
Anas platyrhynchos	3	5.42[a]	5.24–5.65
Heteronetta atricapilla	1	6.15	....
Oxyura jamaicensis	3	6.27[a]	5.93–6.70
Biziura lobata	1	8.71	....

[a] $P < .02$.

TABLE 13

Mechanical Advantage of M. Tibialis Anterior[a]

Species	No. examined	Mean	Range
Anas platyrhynchos	7	.257[b]	.230–.279
Heteronetta atricapilla	2	.250	.234–.265
Oxyura jamaicensis	7	.290[b]	.277–.302
Biziura lobata	8	.296	.270–.332

[a] The force arm is the distance from the proximal articular surface (intercotylar prominence) to the distal end of the tendon scar of the muscle. The resistance arm is the length of the tarsometatarsus from the intercotylar prominence to the tip of the trochlea for digit III.
[b] $P < .001$.

sion of the tarsometatarsus, thus contributing to the actual swimming movement, and they flex the digits. The latter is important in the recovery stroke (tarsal flexion) when he digits are folded together to minimize water resistance, but it is also important in tarsal extension to maintain some degree of digital flexion so as to prevent the water resistance from causing excessive digital extension, which would reduce the effectiveness of the foot as a paddle. The need to keep the digits flexed during the power stroke may explain in part the general increase in size of the digital flexors, but it is difficult to imagine how it might correlate with another phenomenon, the fact that the increased weight of the tarsal extensor (digital flexor) group in *Oxyura* and *Biziura* is greatly augmented by the extreme hypertrophy of a different muscle in each species. These are the proximal head of M. flexor perforatus digiti IV in *Oxyura*, and M. flexor perforans et perforatus digiti III in *Biziura* (table 15). Most likely, these disproportionate increases are simply the most extreme examples in each species of a general trend for enlarged tarsal extensors associated with increased strength of the primary swimming movement in these highly specialized diving ducks.

The recovery stroke of swimming is achieved through flexion of the tarsometatarsus, chiefly by M. tibialis anterior. There does not appear to be any important difference in the relative weights of the two bellies of this muscle, but the total weight of the muscle as a whole increases progressively in the sequence of species illustrating increasing aquatic adaptation (table 12). This adaptation for greater strength of contraction is correlated with an increase in the mechanical advantage of the muscle in *Oxyura* and *Biziura* compared to *Anas* and *Hetero-*

netta (table 13), with *Oxyura* producing 13 percent greater force of contraction compared to *Anas*.

The last muscle to be considered in relation to movements of the foot is M. extensor digitorum longus. As pointed out by Miller (1937:37) it can act as a tarsal flexor if opposed by the toe flexors. Table 15 shows that this muscle is significantly enlarged in the course of aquatic adaptation by the Oxyurini, and it probably has a role in tarsal flexion during the recovery stroke, especially since the digital flexors are active at this time in folding the toes back to reduce water resistance. It must also act in extending the toes at the start of the power stroke.

It is thus clear that increased aquatic locomotion is accompanied by adaptations leading to greater strength of contraction in the muscles primarily concerned with the movement of the paddle through the water, compared to species which are better adapted for walking on land. Presumably this is due to the fact that water resistance is much greater than air resistance because of its higher density.

MOVEMENTS OF THE DIGITS

In addition to the digital extensors and flexors discussed above, there are several intrinsic muscles of the foot which arise on the tarsometatarsus and insert usually at the very base of the individual digits. These muscles, whose functions are indicated by their names, are the adductor digiti II, abductor digiti II, extensor proprius digiti III, abductor digiti IV, extensor brevis digiti IV, extensor hallucis longus, and flexor hallucis brevis. In each species they collectively comprise about 0.40 percent of the dry weight of the leg musculature, and their functions must consist of only the most delicate adjustments and minor movements of the individual digits, for example, in grooming.

The fusion of most of the tendon of M. flexor hallucis longus with that of M. flexor digitorum longus and the degeneration of the small remaining tendon to the hallux in *Biziura* (see above) demonstrates the limited mobility and minor importance of this digit in these highly specialized aquatic birds.

ABDUCTION OF THE LIMB

The gluteus medius et minimus is positioned so as to produce abduction of the limb, but is so small that its action in this capacity must be negligible. The primary usefulness of this muscle is probably in pressing the head of the femur into the acetabulum, and thus contributing to joint stability.

The only effective abductor of the hind limb is the iliotibialis. It is of similar size in *Anas* and *Heteronetta,* but reduced in *Oxyura* and enlarged in *Biziura* (table 15). This muscle must be important in maintaining the abducted position of the limbs characteristic of the swimming posture in these species (see the Summary of Locomotor Habits section, above), and its great development in *Biziura* may be related to the large size of this species. Since the strength of a muscle is a function of its cross-sectional area, a two-dimensional value, while weight is a function of volume, a three-dimensional value, the size of the muscle must be increased more than proportionately in the larger species to perform a similar function in overcoming the weight of the limb during abduction.

TABLE 14

Dry Weights of M. Gastrocnemius, the Tarsal Flexors,[a] and the Sum of Both Expressed as a Percent of the Total Dry Weight of the Muscles of the Hind Limb

Species	No. examined	Gastrocnemius Mean	Gastrocnemius Range	Tarsal flexors Mean	Tarsal flexors Range	Sum of gastrocnemius and flexors Mean	Sum of gastrocnemius and flexors Range
Anas platyrhynchos	3	17.25[b]	16.77–17.82	13.70[c]	13.46–13.94	30.95	30.71–31.28
Heteronetta atricapilla	1	16.47	....	14.54	....	31.01	....
Oxyura jamaicensis	3	18.66[b]	18.02–19.01	19.55[c]	18.90–20.31	38.21	37.86–38.44
Biziura lobata	1	20.13	....	20.58	....	40.71	....

[a] Flexor perforans et perforatus digiti II; flexor perforans et perforatus digiti III; flexor perforatus digiti II; flexor perforatus digiti III; flexor perforatus digiti IV; peroneus longus; flexor hallucis longus; flexor digitorum longus.
[b] $P < .05$.
[c] $P < .001$.

TABLE 15

Weights of the Muscles of the Hind Limb Expressed as a Percentage of the Sum of the Weights[a]

Muscle	Anas platyrhynchos Mean	Anas platyrhynchos Range	Heteronetta atricapilla	Oxyura jamaicensis Mean	Oxyura jamaicensis Range	Biziura lobata
Sartorius	5.80[b]	5.61–6.05	5.95	4.05[b]	3.47–4.92	3.57
Iliotibialis	3.84[c]	3.67–3.97	3.68	2.50[c]	2.45–2.55	6.01
Gas. p. int	8.90[b]	8.86–8.98	10.03	11.26[b]	10.74–11.67	11.27
Gas. p. med	0.57	0.52–0.64	Absent	0.61	0.54–0.68	0.47
Gas. p. ext	7.78[b]	7.35–8.20	6.44	6.80[b]	6.67–6.97	8.39
Ambiens	1.71[b]	1.69–1.74	1.46	1.97[b]	1.91–2.09	0.72
Biceps femoris	6.17[c]	6.00–6.42	5.75	4.51[c]	4.35–4.74	2.71
Glut. med. et min	0.25	0.16–0.31	0.09	0.15	0.07–0.25	0.32
Il. troch. post	4.18[c]	3.91–4.34	3.46	2.66[c]	2.61–2.73	2.28
Il. troch. ant. et med	1.31[b]	1.15–1.50	1.70	2.04[b]	2.00–2.12	1.43
Fem. tib. ext. et med	8.40[c]	7.92–8.76	6.51	5.00[c]	4.85–5.13	5.01
Fem. tib. int	1.25[b]	1.20–1.28	1.05	1.37[b]	1.36–1.38	0.73

Semitendinosus	10.73	10.24–11.11	13.86	9.91	9.44–10.72	13.66
Semimembranosus	1.49[c]	1.42–1.55	1.22	0.62[c]	0.58–0.64	0.42
Piriformis p. il. fem.	4.60	4.42–4.84	5.26	5.17	4.03–6.06	3.94
Piriformis p. caud. fem.	2.51	1.87–3.18	2.03	2.46	2.16–2.77	1.64
Iliacus	0.11[b]	0.06–0.13	0.25	0.25[b]	0.22–0.29	0.01
Ischiofemoralis	1.43[b]	1.37–1.56	1.44	2.25[b]	2.14–2.44	0.79
F. p. et p. d. II	1.91[c]	1.83–2.00	2.26	2.40[c]	2.35–2.44	2.68
F. p. et p. d. III	2.77[b]	2.74–2.83	2.63	3.72[b]	3.48–3.89	6.57
F. p. d. IV (distal)	0.65	0.56–0.69	0.64	0.68	0.60–0.76	0.74
F. p. d. IV (proximal)	2.89[c]	2.53–3.27	3.96	6.93[c]	6.68–7.37	2.94
F. p. d. III (lateral)	0.26	0.21–0.30	0.25	0.28	0.27–0.29	0.73
F. p. d. III (medial)	1.15	1.07–1.22	1.53	1.57	1.30–1.77	1.33
F. p. d. II (lateral)	0.39	0.30–0.49	0.29	0.33	0.32–0.34	0.27
F. p. d. II (medial)	0.20	0.15–0.30	0.34	0.19	0.17–0.21	0.27
Add. long. et. brev.	5.01	4.73–5.23	4.39	4.86	4.38–5.22	2.48
Obturator externus	0.09	0.07–0.12	0.11	0.11	0.08–0.15	0.05
Obturator internus	1.35[b]	1.29–1.44	1.16	1.07[b]	1.04–1.12	0.43
Plantaris	0.34	0.26–0.42	0.26	0.40	0.27–0.51	0.20
Peroneus longus	1.78	1.75–1.82	0.96	1.72	1.46–2.03	3.37
Tibialis anterior (femoral)	1.80	1.72–1.94	1.82	1.85	1.80–1.87	2.94
Tibialis anterior (tibial)	3.62	3.52–3.71	4.33	4.42	4.06–4.83	5.77
Flexor hallucis longus	0.77	0.73–0.80	0.69	0.82	0.75–0.92	0.58
Flexor digitorum longus	0.93	0.85–1.04	0.99	0.91	0.79–1.06	1.10
Popliteus	0.12	0.10–0.13	0.14	0.14	0.06–0.21	0.14
Peroneus brevis	0.41	0.36–0.44	0.36	0.52	0.49–0.53	0.59
Extensor digitorum longus	1.98[c]	1.88–2.06	2.35	3.02[c]	2.97–3.08	3.00
Ext. pro. dig. III	0.02	0.01–0.03	0.02	0.03	0.02–0.03	0.06
Ext. brev. dig. IV	0.09	0.08–0.09	0.07	0.07	0.06–0.08	0.10
Abductor digiti II	0.09	0.06–0.11	0.09	0.07	0.07–0.08	0.15
Extensor hallucis longus	0.04[b]	0.03–0.05	0.05	0.01[b]	0.01–0.02	0.01
Abductor digiti IV	0.14	0.13–0.16	0.14	0.13	0.12–0.13	0.08
Abductor digiti II	0.03[b]	0.02–0.05	0.03	0.09[b]	0.07–0.10	0.05

[a] Based on three specimens each of *Anas* and *Oxyura*, and one each of *Heteronetta* and *Biziura*.
[b] $P < .02$.
[c] $P < .001$.

ADDUCTION OF THE LIMB

There are no muscles specifically positioned to act as adductors of the limb. The adductor longus et brevis, despite its name, functions primarily as an extensor of the femur. Adduction of the limb is apparently produced by the action of several muscles whose primary lines of action in other directions have components which will cause adduction. The sartorius, ambiens, adductor longus et brevis, semimembranosus, and iliacus apparently serve in this way. Contraction of any one of these when the limb is abducted would produce some movement additional to adduction, for example, the sartorius would cause adduction, com-

TABLE 16
RATIO OF THE PERCENT-WEIGHT OF THE TARSAL FLEXORS[a] TO THE PERCENT-WEIGHT OF M. GASTROCNEMIUS

Species	No. examined	Mean	Range
Anas platyrhynchos	3	0.80	0.76–0.83
Heteronetta atricapilla	1	0.88	....
Oxyura jamaicensis	3	1.05	1.00–1.13
Biziura lobata	1	1.02	....

[a] Listed in table 14.

bined with flexion of the femur. Actual adduction is presumably performed by some balanced combination of these muscles so as to eliminate any unwanted secondary movements. A large force is not needed for adduction since the movement is aided by gravity.

GENERAL CONCLUSIONS

Heteronetta is clearly intermediate in many respects between *Anas* and the advanced stifftails in the osteology and myology of the tail and hind limb. Except for the pelvic girdle, whose general form is very much like that of *Oxyura*, most of the characteristics are qualitatively more like those in *Anas* than in the other stifftails. This is especially true in the musculature of the hind limb, which has a number of highly specialized features in *Oxyura* and *Biziura*, such as the modified character of Mm. flexor perforans et perforatus digiti II, gastrocnemius pars interna, and most notably, the adductor longus et brevis, which are lacking in *Heteronetta*. Furthermore, the advanced stifftails have developed marked changes compared to *Anas* in the lever actions and relative sizes of certain muscles, all of which are parts of a complex of changes converting a semi-terrestrial to a fully aquatic limb design. In these features *Heteronetta* is usually either similar to *Anas* or intermediate between it and *Oxyura*. Thus the results of this study clearly support the opinion that *Heteronetta* is a connecting link between *Anas* and the other stifftails, and represents an early, not very advanced, stage in the evolution of the more specialized types. Of course it is not the direct ancestor of any living genus. This is shown by the presence of certain unique features which have evolved since the separation from its common ancestor with the other genera, such as the loss of M. gastrocnemius pars media, and the development of fully parasitic nesting behavior, yet anatomically it cannot have diverged very far from this common ancestor.

The above data give rise to a question as to the proper taxonomic position of *Heteronetta*. In many ways it resembles *Anas*, including the form of the bill and plumage of the downy young, the unlobed hallux, some aspects of sexual behavior, the structure of the cnemial crest and many aspects of the myology, as described in detail above. For these reasons one might consider placing it in the tribe Anatini along with *Anas*. This would leave the Oxyurini a more narrowly defined, cohesive group, since *Oxyura* and *Biziura* are much more similar to each other than either is to *Anas*. But such an arrangement would fail to make explicit the obvious relationship which *Heteronetta* has with the other genera, and which is demonstrated by so many diverse characters of anatomy and behavior. For this reason, the present classification should be retained, but with the understanding that *Oxyura* and *Biziura* represent extremes of specialization which are only moderately developed in *Heteronetta*.

In terms of their aquatic locomotor systems, *Oxyura* and *Biziura* represent two end products of a long, and in part, common history of development of highly specialized diving forms from a non-diving ancestral stock much like *Anas*. This has involved a complex of skeletal-muscular modifications, including the relative proportions of the pelvic girdle and hind limb, and the relative sizes, lever actions, and structural modification of muscles of the hind limb and tail. These are all components of one great adaptive change leading to increased efficiency of aquatic locomotion at the expense of efficiency in terrestrial locomotion.

There is evidence that *Oxyura* and *Biziura* shared a common ancestry for some time after diverging from the line which gave rise to *Heteronetta*, but that they diverged from one another before either reached its present extreme stage of specialization. Certain shared characters must have arisen after divergence from *Heteronetta* but before the separation into the two living genera. These include the specialized form of the gastrocnemius pars interna and adductor longus et brevis, and partial reduction of the patellar head of M. flexor perforans et perforatus digiti II. It is possible that at the time of separation these characters were not as highly developed as at present, and reached their extreme condition in a separate but parallel fashion, but it seems unlikely that they would have arisen in so similar a fashion entirely independently. Other specializations, however, apparently arose after divergence of the two groups. Thus in the caudal musculature the extreme hypertrophy of M. levator coccygis and levator caudae in *Biziura* is not attained in *Oxyura*. In the hind limb there has been extreme hypertrophy of one of the digital flexors, apparently for increased strength of tarsal extension, but it is a different muscle in each genus which undergoes this change.

Although both *Biziura* and *Oxyura* are highly specialized diving ducks, it appears clear that *Biziura* is the most highly specialized of the Oxyurini in its locomotor, and other, adaptations. In many features which have been regarded as specializations for aquatic locomotion, *Biziura* shows more extreme development than *Oxyura*. These include the structure and proportions of the pelvic girdle, the length of the tail, the relative sizes of the caudal muscles, the proportions of the hind limb and structure of the patella, and the relative sizes and mechanical advantage of various muscles of the hind limb. Extreme specialization is also found in the remarkable sexual dimorphism and sexual behavior (Johnsgard, 1966) of the species. All of these specializations define a very limited ecological niche, and *Biziura* is limited to a single species on one continent. Probably it achieved its specialized condition in Australia, and its present poor flying abilities prevent its movement to other continents. *Oxyura*, somewhat less specialized, has managed to colonize much of the world with six, usually allopatric species (Delacour, 1959: 225). *Heteronetta,* the least specialized of the Oxyurini, might thus be expected to be the most successful, yet like *Biziura* it is limited to a single species of limited geographical range. Probably this is because the potential ecological niches which it might occupy elsewhere are already more effectively utilized by members of the other diverse and successful groups of diving ducks which have evolved separately from an ancestral group similar to that which gave rise to the Oxyurini.

SUMMARY

The tribe Oxyurini includes three genera of ducks which show progressive modification for increased swimming and diving ability. The group evolved from an ancestral form similar to the genus *Anas* of the tribe Anatini. *Heteronetta* is intermediate in form between *Anas* and the more highly specialized genera *Oxyura* and *Biziura*. Various modifications of the osteology and myology of the hind limb and tail have occurred which improve the efficiency of an adducted leg posture in diving, and the use of the tail as an underwater rudder. These include lengthening of the tail and enlargement of the caudal levator muscles, narrowing of the pelvis and elongation of the postacetabular portion, enlargement of the area of origin of leg muscles from the knee area, reduction of the size of thigh muscles and increase in shank muscles correlated with the change from walking to swimming. Changes in the line of action of certain thigh muscles improve their effectiveness as fixators of the thigh during diving. An increase in the mechanical advantage of many muscles may be associated with the need for strength of action rather than speed, in swimming as compared to walking.

ACKNOWLEDGMENTS

I am grateful to Dr. Peter L. Ames and Dr. Wilbur Quay for their encouragement and guidance during the course of this study. Dr. Joseph T. Gregory offered many helpful criticisms of the manuscript. James Lynch, Paul Covel, William Arvey, and John Cowan helped me to obtain specimens of *Anas platyrhynchos* and *Oxyura jamaicensis* for this study. I also wish to thank Dr. Charles G. Sibley and Eleanor Stickney of the Peabody Museum of Natural History, Yale University, and Dr. Robert W. Storer of the University of Michigan Museum of Zoology, for the loan of specimens in their care. Gene M. Christman gave me much helpful advice on the preparation of the illustrations.

LITERATURE CITED

ALEXANDER, R. MC N.
    1968. Animal Mechanics. Seattle: Univ. Washington Press.

BELLAIRS, A. d'A., and C. R. JENKIN
    1960. The skeleton of birds. Ch. VII in A. J. Marshall, ed., Biology and Comparative Physiology of Birds, vol. 1. New York and London: Academic Press.

BOAS, J. E. V.
    1933. Kreuzbein, Becken und Plexus Lumbosacralis der Vogel. Danske Videnskabots Skrifter, Naturvidenskabelig Ser. 9, 5:3–74.

DABELOW, A.
    1925. Die Schwimmanpassung der Vögel. Ein Beitrag zur biologischen Anatomie der Fortbewegung. Morph. Jahrb., 54:288–321.

DELACOUR, J.
    1959. The Waterfowl of the World, vol. 3. London: Country Life Ltd.

DELACOUR, J., and E. MAYR
    1945. The Family Anatidae. Wilson Bull., 57:3–55.

DU TOIT, P. J.
    1913. Untersuchungen uber das Synsacrum und den Schwanz von *Gallus domesticus* nebst Beobachtungen uber Schwanzlosigkeit bei Kaulhuhnern. Jenaische Zeitschr. Naturw. 49:149–312.

FISHER, H. I.
    1946. Adaptations and Comparative Anatomy of the Locomotor Apparatus of New World Vultures. Amer. Midl. Nat., 35:545–727.

FORBES, W. A.
    1882. Note on some points in the Anatomy of an Australian Duck (*Biziura lobata*). Proc. Zool. Soc. London, 1882:455–458.

FRANTISEK, J.
    1934. Les Muscles de l'extrémité posterieux chez l'oie et chez le canard. Biol. Spizy veterin. Brno, 13 (no. 179):1–20 (1934) n.v.

FRITH, H. J.
    1967. Waterfowl in Australia. Sidney: Angus & Robertson.

GEORGE, J. C., and A. J. BERGER
    1966. Avian Myology. New York and London: Academic Press.

HOWARD, H.
    1929. The Avifauna of Emeryville Shellmound. Berkeley: Univ. Calif. Publ. Zool., 32:301–394.
    1950. Fossil Evidence of Avian Evolution. Ibis, 92:1–21.

HUMPHREY, P. S., and G. A. CLARK, JR.
    1964. The Anatomy of Waterfowl. Ch. IX in J. Delacour, The Waterfowl of the World, vol. 4. London: Country Life Ltd.

JOHNSGARD, P. A.
    1965. Handbook of Waterfowl Behavior. Ithaca: Cornell University Press.
    1966. Behavior of the Australian Musk Duck and Blue-Billed Duck. Auk, 83:98–110.
    1968. Waterfowl, their Biology and Natural History. Lincoln: Univ. Nebraska Press.

KORTRIGHT, F. H.
    1942. The Ducks, Geese, and Swans of North America. Washington, D.C.: Amer. Wildlife Inst.

MILLER, A. H.
    1937. Structural Modifications in the Hawaiian Goose (*Nesochen sandvicensis*). A study in Adaptive Evolution. Berkeley: Univ. Calif. Publ. Zool., 42:1-80.

OWRE, O. T.
    1967. Adaptations for locomotion and feeding in the Anhinga and the Double-crested Cormorant. Ornithological Monographs, no. 6.

PYCRAFT, W. P.
  1906. Notes on a skeleton of the Musk-duck, *Biziura lobata* with special reference to skeletal characters evolved in relation to the diving habits of this bird. Jour. Linn. Soc., Zool., 29:396–407.

QUENNERSTEDT, A.
  1872. Studier I Foglarnas Anatomi. Acta. Univ. Lund., 9:1–61.

REYNOLDS, S. H.
  1913. The Vertebrate Skeleton. Cambridge, Eng.

RUCK, P. R.
  1949. Studies on the Muscles of the Pelvic Appendage in Certain Anseriform Birds. Unpublished M.S. thesis. Pullman: State College of Washington.

SHUFELDT, R. W.
  1909. Osteology of Birds. New York State Mus. Bull. 130.

STOLPE, M.
  1932. Physiologisch-antomische Untersuchungen uber die hintere Extremitat der Vogel. Jour. f. Ornith., 80:161–247.

STORER, R. W.
  1960. Adaptive Radiation in Birds. Ch. II in A. J. Marshall, ed., Biology and Comparative Physiology of Birds, vol. 1. New York and London: Academic Press.

VERHEYEN, RENÉ
  1955. La Systematique des Anseriformes Basée sur l'Osteologie Comparé (suite). Bull. Inst. Roy. Sci. Nat. Belgique, vol. 31, no. 35:1–18; vol. 31, no. 36:1–16; no. 37:1–22; no. 38:1–16.

WEIDMANN, U.
  1956. Verhaltensstudien an der Stockente (*Anas platyrhynchos* L.). Zeit. fur Tierpsychol. 13: 208–271.

WELLER. M. W.
  1967. Notes on Plumages and Weights of the Black-Headed Duck, *Heteronetta atricapilla*. Condor, 69(2):133–145.
  1968. The Breeding Biology of the Parasitic Black-Headed Duck. The Living Bird, vol. VII: 169–207.

WETMORE, A.
  1926. Observations on the Birds of Argentina, Paraguay, Uruguay, and Chile. U. S. Nat. Mus. Bull., 133:1–488.
  1960. A Classification for the Birds of the World. Smithsonian Misc. Coll., vol. 139, no. 11.

WOOLFENDEN, G. E.
  1961. Postcranial Osteology of the Waterfowl. Bull. Fla. State Mus., vol. 6, no. 1:1–129.